Behind the Scenes

Behind the Scenes

Cora Steyermark

Missouri Botanical Garden

To
Julian Steyermark,
the X.Y: hero of the story

The botanical nomenclature, both scientific and vernacular, in this book follows that of Julian A. Steyermark, *Flora of Missouri,* Iowa State University Press (1963). Vernacular names of plants begin with uppercase letters, while those of animals have been lowercased.

The cover illustration is based on a watercolor of *Liatris aspera,* executed by Julian A. Steyermark in August 1927.

ISBN 0-915279-02-9

Library of Congress Catalog Card Number
84-60643

CONTENTS

INTRODUCTION

The purpose of this story is to give the reader an insight behind the scenes during the preparation of a flora and to reveal to him the research of which he seldom guesses.

Over a period of 30 years the reader will follow the author of the flora, indicated here as X.Y., from his home in Illinois to Missouri and over thousands of miles within the state by car during his research for the 1,728 page "Flora of Missouri."

They will follow him every month from March to November on foot hundreds of miles through rain, scorching sunshine, uphill and down, over rivers and through pathless forests in his determination to ferret out every hidden plant in the state.

They will experience the pleasures and miseries of his camps; problems encountered; thrilling surprises in plant life; landscape jewels never seen before by anyone capable of appreciating such canyons, or moss-covered and flower-covered slopes far from trails and highways. And they will meet friendly and helpful people.

It may seem that the story of the "Flora of Missouri" would be severely slanted to that one state. However, the flowers and birds discussed are common also to all central and northeastern states; the folklore; the conservation problems; and the sentiments and concepts are common to people all over the world.

It is the intention here to share some of the many experiences as the flora develops behind the scenes and finally comes to completion.

I, the X.X., went along to take notes and care for the camp.

CHAPTER I

MARCH

Spring or Winter

For three days the sun had warmed the soil enough to bring up the Harbinger of Spring (*Erigenia bulbosa*), a pea-sized head of mixed salt-and-pepper–like flowers casually passed up as a splash of frost on a bit of weathered leaf. A few Bluets suddenly appeared in open grassy spots. Bluebirds busily inspected holes and hollows repeating the song that never fails to set me wondering why a song heralding new life and gladness should have tucked into every syllable a tone of querulous sadness. Such a preview of spring during this particular year led us to believe the season was earlier than usual in the Ozarks. With every sign pointing to a balmy weekend, on March twelfth we packed for the first expedition of the year—ready to start soon after daybreak on the following morning.

At three o'clock on that particular morning of March thirteenth, while at a small town in Missouri, I was awakened by rain driving against my bedroom window. Dismissing it as a good spring rain to hasten the wildflowers, I went to sleep again. By daylight the rain had turned to sleet and ice glazing the ground with a blinding glare. At breakfast we debated the wisdom of driving out on ice. We were still debating to the last cup of coffee, when nature decided for us. Sleet changed to large snowflakes which fell heavily, capping the ice two inches by nine o'clock. That seemed enough to keep the car on the road. Slip chains were buckled over the tires, and we started out with the thought in mind that from warm rain, to sleet, to snow in five hours, it could as well go to warm sunshine and bare ground before the day ended.

A stiff wind driving the heavy snow at a forty-five degree angle added inches, as we added miles through the white silent woods. We were the only ones boldly venturing over the miles of unbroken road. We knew the road was there, because the trees and shrubs were not. On top of a hill halfway to Westover Springs in Crawford County, X.Y. discovered he had lost a slip chain. After much slipping and sliding, and the loss of a second chain, the fourteen miles of unbroken road were traversed, and our first goal was reached.

For several years past, X.Y. had been surveying Missouri springs collecting data for a paper on the flora of freshwater springs, just one paper sandwiched in between research on the big "Flora." Westover Spring rates among those of second magnitude in Missouri. It rises out of the ground in a level region, covers an area of about two acres of marshy lake, curves around an island, then unites into a broad stream containing Water Cress, Water Milfoil, Bur-reed, Water Starwort, Pondweed, and many other water plants.

In former times part of the water had been diverted through a millrace to turn an overshot wheel for power to grind grain of neighboring farmers. When the mill ceased to run, the race was converted for trout propagation.

Now while X.Y. fished for plants, I watched the birds. That a fish race could become a choice bit of feeding ground for birds never had occurred to me. While snow still fell heavily, hundreds of birds were at work. The fish had been fed ground meat that morning. Pieces of meat thrown to the sides of the race by lashing hordes of trout were grabbed up by the birds and eaten either in the shelter of the banks or dropped on the snow, leaving thumb-sized blood stains dotting the banks throughout the race length.

Juncos, song sparrows, robins, cardinals, starlings, and bronzed grackle were most abundant. We were told that a flock of approximately 1,000 robins had visited the race the day before. Flying crows looked gray through the slanting snow; and the notes of bluebirds in the trees overhead made a strange medley with the soft patter of snow through branches. On floating islands of water plants birds kept up a constant feeding. Snowbirds lit on a clump of Cress, hastily picked up something, either a salad to finish off the menu of meat from the race below, or tiny black periwinkles, or perhaps lodged Sycamore seeds; sang their fairy-like tinkle; and merrily dashed to another clump. Here, also song sparrows and tree sparrows, feeding in a similar manner, often sank until half the breast was submerged. The temperature was gradually dropping, and I wondered what would prevent their wet feathers from freezing when they left the warm spring water.

Water plant specimens which X.Y. had collected froze in the air and snapped off. By noon the radio in our car announced 28 degrees. The snow was now four inches deep and still coming down.

Goldfinches overhead announced their presence with contented, "cheers, tcheers," between the showers of Sycamore seeds which they dislodged. Killdeers along the creek sent out their notes of alarm. Red-winged blackbirds which had been in the country since March seventh called, "chonk-err-ee, chunk-o-meat," on their way to the fish race. Near the source of the spring, three Canada geese, breast-deep in snow, walked disconsolately through the low, bare undercover.

All birds seemed to be incautious in their hurry to get food. Often we walked to within a few feet of the feeding birds. A downy woodpecker was so busy pecking a hole in a vine that he scarcely noticed us until we could almost touch him before he decided to leave the worm he was so determined to get. The vine, about an inch in diameter, had six holes freshly drilled between the ground and lower limbs of the tree to which it clung.

Leaving this busy scene, we headed toward Cold Springs in Washington County some 50 miles distant. On the road we noted birds feeding along slender threads of water at places where the difference in potential afforded enough energy to keep it from freezing.

Near Potosi we left the highway to follow the Hillendale Forestry Road with its one foot of fresh, unbroken snow. At higher elevations along ridges, only an occasional titmouse or chickadee ventures out on ice-covered limbs for no other purpose, that I could see, than to listen to the tinkle when one limb, pendent with inch-long icicles, swayed against another similarly laden. Brittle was the dominant theme along this isolated road. Trees creaked and crackled with their icy load, and the car tires shrieked in the dry snow. Another slip chain had departed silently from the creaking tires when no one was aware.

At Cold Springs by four o'clock the temperature had dropped fifteen degrees lower. How appropriately the name "Cold Springs" fitted on this day. Snow, one foot on the level and boot-top deep in drifts, was now coming in stinging gusts of fine snow. But the birds were feeding on floating water plants as at Westover. At the head of the spring I saw the first towhee of the season fly from a shelter of overhanging bank to a little dry cave under a root outcrop over the branch. In the small race leading from the spring, a junco made a dive for something and came out with a drowned angle worm. Flying to a sheltered cavelet, he ate the washed out worm with as much relish as a robin eats a fresh one.

Guarding our ears from the stinging-hard granules of snow, we stumbled through drifts from one open pool of water to another until each was inspected.

And now the time had come to turn back, yet X.Y., as usual, gave up reluctantly. He must stop in the wooded creek bottom to look for flowers of Ozark Witch Hazel. In early dusk he found them—each orange-yellow petal now spread out like a miniature octopus with half its arms stiff in an icy cast. He cut a branch for me, first warming cold hands inside his jacket to free the flowers from ice so I could smell the spicy fragrance.

"Smells like the witch hazel shaving lotion you once used," I said.

X.Y. related its virtues, "Magic water the Indians called the astringent remedy they brewed from stems and leaves for sore muscles, nose bleed, and eye diseases. Early white settlers took up the practice and passed it down to the present day lotions which serve for little more than a mental boost."

Repeated crossings of the winding creek added to our load. Water froze immediately on wheels and under parts of the car forming large chunks of ice. By now all chains had slipped away and we were on our own. As we left the creek bottom and started up a steep hill, I looked longingly at a small white house by the road where the light of a kerosene lamp showed through drawn shades. I sniffed at the wood smoke curling blue above the snow-covered roof and grew colder by contrast with the warmth I imagined inside that house.

We had miles of snow-covered roads between us and a place to spend the night. The only house we had seen since two o'clock was this one sitting so snugly at the foot of the hill. I was cold and had been all day. I was still looking back at the house when one-third of the way up the hill, the car began a side glide and stalled. X.Y. shut off the motor. "Get out," he said, "there's a deep ditch on my left, and I'm within inches."

I got out and floundered in deep snow, hunting for something to block the rear wheels. An American woodcock, startled by a shower of snow and ice hurled into its shelter by the spinning wheels of our balked car, started headlong down the steep hill through the bushes. I stood watching his clumsy descent.

"What are you looking at?" The tense voice of X.Y. brought me back to our predicament.

"That short-legged woodcock stumbling down the hill. With no neck and a bill three times as long as his tail, I think he could have done better running backwards. I expected him to somersault anytime."

"Suppose you somersault down to that house for help, X.X. It's no use—I can't make this hill."

I went down, knee-deep in snow, tried the front gate but failed. The gate leading to the back door was open. At my knock on the kitchen door a woman answered. No, the men, her husband and brother, could not help. They had no horses, no truck, or tractor. They were not farmers. The men had just taken warm baths, and her brother had a cold. No, there was no place in walking distance where one could spend the night.

Strangers, whoever or whatever they might be, could not be permitted to sit all night in a car with the temperature near zero. We were invited in to share the best they could offer.

Our cold left-over lunch was brought to the kitchen, already chilled since the fire from the wood-burning cook stove had been permitted to die out. A circular ring on the kitchen floor—evidence of the baths in the family washtub—sent further chills up and down my spine. We wondered what the family in the living room were conjecturing about the strangers in their kitchen.

After we had finished our hasty supper we sat for two hours by a red-hot heating stove in the living room trying to invent some subject of common interest and to put them at ease about our presence. Finally, the three-year-old girl asleep across her mother's lap was carried to a cold bedroom. I was given the small, unheated bedroom with a single bed, Mrs. A's brother's room. Snow had filtered in through a broken window pane. With boots off for comfort and my coat for an extra cover, I slept fairly well. X.Y., on the floor of the living room on one side of the red-hot king heater and Mrs. A's brother on the other side of the heater, said they roasted all night.

The next morning we were treated to a breakfast of hot biscuits, ham and eggs, fruit, and honey—and men ready to help us get up the hill. The little girl, with inhibitions less socialized than her parents, accepted the money we offered for their trouble and courtesy.

The morning was cold and clear. Pine trees and shrubs, gray and bent with ice and snow, lent variety to the icicled oaks. At the top of the hill we came upon the sunrise. The whole view for miles across the hills was a brilliant blaze of crystalline purity. Against the sun, the tops of trees turned to polished gold, shading to silvered lower limbs rising above a soft setting of snow-covered shrubs.

X.Y. stopped the car. "I'm stuck with a view," he said. "I'm afraid I'll break something if I move."

"That whole view looks as if it might break if you breathed on it. This is something you see once in a lifetime. I'm glad we stalled last night," I said and really meant it.

A chickadee lit on an icy limb at our right, slipped to an upside down position, but hung on. We were reluctant to break the spell and move on, but more springs had to be visited. Along the highways doves cooed from fence rows and telephone wires. A marsh hawk sailed out across the field.

The load of ice we carried under the car and the unbroken roads kept our speed to a minimum, making the distance seem three times as great to the next three springs we expected to visit.

At Beaver Spring in Crawford County we found the same species of birds feeding on water plants. Here also phoebes were feeding. From a perch on a post near the edge of the lake formed by the spring, they made frequent dives at the water for something which was totally invisible to me from the same distance. A whistling swan opened his wings to the wind and sailed near the shore toward me, retreating to the opposite shore and returning to the same place several times. Two teals swimming on the lake flew before identified. All around the lake a lacework of small bird tracks covered a width of 15 feet from the edge.

The call of the flicker led us through thickets of Sycamore, Ironwood, and other shrubs along the spring branch. Dozens of last year's nests piled high with mounds of snow showed this thicket to be a favorite nesting site. Here the winter wren was seen trying to get a meager meal from something on exposed roots and twigs in little coves along the edge of the water. Horned larks twittered in rapid flight overhead, throwing shadows on the field of snow, rippled into east and west zigzags by the north wind.

The late afternoon sun had done nothing in the way of thawing, but the 12 miles of highway leading to Meramec Spring had been scraped. Along this graveled highway, ten flocks of from 50 to 100 juncos each were flushed. Small gravel on the highway was the object of attraction.

At Meramec Spring were the same species of birds, juncos, song sparrows, and tree sparrows, feeding on floating plants. Phoebes and bank swallows make their homes in round cavities under the bluff overhanging this large, bottomless spring, one of Missouri's largest. Over the main body of this spring where no plants grow, phoebes darted low, poised one-half foot above the water, motionless except for vibrating wings, for thirty seconds or more, then after a sudden dip at the water rose several feet into the air, and then repeated the act. They were catching insects, of course, but I could not see them. From the opposite hill came the call of the Carolina wren. And far down along the wooded banks of the spring branch, the kingfisher announced his watch over the water.

Meramec Spring in Phelps County, six miles southeast of St. James, is the sixth largest spring in Missouri. Boiling up from under a bluff with an average flow of 96,000,000 gallons and a maximum flow of 420,000,000 gallons of water per day, it was at one time considered as the potential water supply for the city of St. Louis. One mile downstream, the spring branch flows into the Meramec River swelling it to twice its previous size.

While X.Y. hastened to study water plants here, I walked to the last symbol of the old ironworks, an open-hearth furnace of cut stone still standing in front of the mountain of slag which passed many years ago through its bowels.

In 1826 Thomas James erected the furnace, the first really successful attempt to mine the iron ore of Missouri. The ore was dug from the slopes one-half mile west and hauled by wagon downhill to the smelting furnace. Repeated droppings of red ore left the roadside red to the present. Virgin timber was cut for miles around to build coal pits, miniature volcanoes which sent smoke curling from top and sides as the green wood smouldering without blaze slowly changed to charcoal to feed the smelting furnace. For years smoke curled upward all over the nearby counties. Only a few virgin trees still exist in vast areas of secondary growth near smelting furnaces.

Pig iron "pigs" produced in the blast furnance were shipped by wagon train to St. Louis, something over 100 miles away, until 1859 when the St. Louis–San Francisco Railway was constructed through St. James, only six miles from the furnace. In 1873, the depression closed the furnaces.

Large durable stones used to dam the spring still remain, likewise the millrace which carried some of the spring water to turn a huge overshot water wheel furnishing power to grind the grain of farmers. A farmer with a sack of grain to be ground threw the sack across the back of a horse and rode to the mill, had the grain ground and carried his grist home—a familiar sight until 1900, when the old moss-covered wheel ceased to turn for profit. For years it turned on—a rare picture of clear, blue water and cool, green moss. The old wheel grew softer and slower as moss piled more thickly until finally blade by blade it fell apart.

I looked about for some "pigs" which used to lie around, but none remained. At one time some one tried to sink one in the spring to test its depth, but the force of the water pushed the heavy iron away.

I stood alone on this once bustling spot where thumping grist mill, noisy blast furnace, and hurrying workmen ground on for 60 years; and now it was back to the quiet of nature—not a soul in sight, not a sound of man-made devices. Only the stony, snow-covered blast chimney backed by its hill of slag remained as testimony to the ambitions of Thomas James. Though the man and his enterprise had passed, he had supplied a useful link in the progress of a nation.

The twittering of birds overhead in a giant Sycamore brought me back to the present. I looked up to sort out the various bird notes. Many goldfinches in their olive-green winter coats swung like acrobats on olive-brown balls dangling by the hundreds from greenish-gray limbs. A dozen or more purple finches also fed on the Sycamore balls. Among shrubs at the base of the trees, titmice and chickadees scolded, picking up seeds dislodged from the banquet above. This dainty picture, perfectly blended in quaker-like colors, made a pleasing contrast to the gypsy scene of the cardinal posing on clean snow against the bright green moss-covered rocks kept evergreen by a spray from the remaining falls of the old dam, a most enduring relic of the old ironworks.

The sun was down, and snow again slanted across cold, bare fields as we ended the day with a visit to Evans Spring. A covey of quail moved from their shelter as we entered the woods. On this spring and its branches, the same species of birds fed as on the four preceding springs, though X.Y. had some variations in plants. Near dusk, when we left Evans Spring, the birds were still feeding.

So ended the two days with twenty-eight species of birds observed braving the snow storm for food. And we thought spring had come.

CHAPTER II

APRIL

Making Specimens

Following the regular day's work at offices in Chicago and a 350 mile drive, we arrived again in Missouri on April 20. This time we came for further research on X.Y.'s proposed "Flora of Missouri."

On Brazeau Creek in Perry County, where we had found a nice place to camp, we went hiking after dark. X.Y. was so anxious to feel Missouri dirt under his feet he could not wait until daylight the following day. With flashlights to guide us, we crossed the creek at our camp, using projecting stones to avoid the water. After crossing a field, we skirted along the low limestone ledges looking at Larkspur and Columbine inaccessible to cattle. I wondered how their appealing blues and reds had escaped the hands of people. Finally we climbed the bluffs. From the top of the highest bluff a hazy moonlit view spread out below us. The view was enchanting, but a plant hunt by flashlight yielded nothing. We returned to our camp across a field of oats nearly knee-high and wet with dew.

After the long drive and this initial hike for the trip, we slept soundly until the birds roused us early the next morning. Twenty-three kinds of birds were seen and heard during breakfast. Floating from tree limbs over the nearby creek came the elusive, modulated song, fresh from the throat of the Louisiana waterthrush. These phrases of tranquility, occasionally accompanied by the rat-a-tat-tat of the pileated woodpecker calling attention to his find of a dead tree, provided a two-phase aspect of honey and pepper for breakfast.

As we saw children starting to school we were reminded of our tardiness and made haste to break camp and drive to an old field on top of a Mississippi bluff where we left the car, shouldered our loads, and set out over the bluffs so shut in by hills on all sides that any cooling breeze that might venture up a ravine was diffused with no more evidence than the trembling leaves over our heads. Already the sun was proclaiming undisputed almightiness, and the knapsack of lunch and notebooks I carried were burdensome until I took a look at X.Y. with his cameras, huge vasculum, maps and notebooks, then I considered my load light.

We trudged on with increasing weariness and hunger. With each successive day of exercise this weariness would decrease. Hills could be climbed with fewer stops and less puffing. Our office-softened muscles would grow firmer and more useful. At an interesting place the hour hands of our watches reached the perpendicular. X.Y. said, "Halt," and we dropped our loads. During lunch we sat on the edge of a promontory commanding a view of wide stretches of Illinois farmland beginning to come to life across the river. An occasional tantalizing breeze coming up from the cooler valley below passed over our perspiring faces temporarily giving blissful relief. Below us ferns, Columbines, and Shooting Stars softened the precipitous rocks completely, beyond our reach. After lunch we hiked downhill to a ravine where a waterfall ended our descent to the river. As we climbed the next steep hill counting white and lavender members of Shooting Star colonies, I counted seven tortoises out walking on the sunny hillside and one blue racer snake. X.Y. found early Coral Root (*Corallorhiza Wisteriana*) which was new to me, but the Azaleas we hoped to find were not in evidence.

The Azalea (*Rhododendron roseum*) occurs in 15 southeastern Missouri counties. One white form is found in Ste. Genevieve County. This waist-to-shoulder high, rather open-branched shrub normally grows on acid soils over sandstone, chert, or granite in semishade, however it seems to prosper in open fields where farmers have cut away the timber and cattle have pruned the plant. There the densely-branched shrub crowds the lovely, fragrant pink flowers into showy masses.

When searching for plants X.Y. always carried topographical, section, and contour maps to keep track of the location of plants he collected, and incidentally ourselves when we strayed too far from any special landmark. We were not limited to paths, or trails, but struck out through woods, thickets, down or upstream, or across hills and hollows. I would sometimes leave behind a cap, bandeau, or scarf on a limb or bush which had snatched off the head covering, rather than retrace steps over the troublesome area and risk getting behind. For, if once I got too far behind to call or hear an answer to my "yohoo," I would have to return to the car since I never knew how far he would go in that direction, or where he would turn to the right, or the left. He, himself, was guided by the general condition of an area when he reached it, and such a thing as a planned march was not possible. It was up to me to keep up or find my way back alone to the car or campsite.

Now having reached the top of the second ridge we scrambled down the hill and again found the ravine too steep to get to the river. Up the next hill we went, and down and up many more before a place was found for a descent to the bottom. What a relief it was to reach level ground after a day of slipping, sliding, perspiring, and climbing each hill just to go down to the other side, scramble over boulders, tree trunks, and puddles in ravines, just to go up hill again. X.Y.'s guiding aim put wings to his heels, while my lowly aim was to find a safe place to put my feet and keep up.

At last we relaxed on level bottomland and walked through beds of *Trillium flexipes* and Wild Leek (*Allium tricoccum*), which in April has only broad leaves resembling those of the tulip. The young leaves can be used for seasoning, soups, or in a green salad. At this time we were not in the mood or need of either. In June the leaves of this Wild Leek die down and the flower stalks rise to complete the cycle. X.Y. was thrilled at finding Leek in this area. This unexpected find lifted him above the border of exhaustion. In comes a new cross for a county record, out goes fatigue or dejection.

The return along the railroad tracks was a respite for the long steep climb uphill to the car where we arrived at five o'clock, hot and dry. By this time a storm was coming up.

Rain had already muddied side roads to Apple Creek where we stopped at a spring branch. X.Y. still had enough energy to examine the surrounding hills before dark, while I cooked supper in a cold drizzle on a gravel bar. A man who is capable of persisting over and above the average modicum is equipped either with singlemindedness, or an exceptionally robust vitality. X.Y. was endowed with both. His short expedition was not in vain as on the nearby hills he had found the anticipated Azaleas.

After we had eaten supper in the car during the rain, X.Y. had the difficult task of putting the specimens collected during the day into press. He let me see and smell the sprigs of Azalea which seemed too lovely to pick and mash flat in a press.

Putting plants into press at night was not the most interesting phase of plant collecting. Beginning with the last number in his notebook, a number now running into sixty thousand, he would add the next number, then put the same number on the blank edge of a folded sheet of newspaper, place the specimen on the sheet, and arrange the flowers and leaves so that both upper and lower surfaces would appear when the specimen was dried. Also the root and fruits were added, if they had been obtained.

Having added the specimen to the press and a blotter for absorbing the moisture, he then wrote the name of the plant, if known, or left a blank following the number, if the plant had to be identified. Then he followed the number and name, or blank in the notebook with all the information observed about the plant: the date, the approximate distance to the nearest town and highway, county, or section of the county, and details of habitat such as n-, s-, e-, or w-facing hillside, bottomland, field, woods, and type of soil where the plant was growing. Also notes were added on the color of flowers or leaves, pubescence, or bloom on stems, or any other characteristics that might be lost in drying.

Then the next number followed in the notebook, and the next plant specimen added to the press. This went on in a series until every specimen was under press. If collecting had been good that day, the number might be

100 or more. Also one to several duplicates were made of each specimen. Then straps were tightened on the press, and the day's work was done, unless he had to work on the county maps.

After the plants were put into press and the rain had ceased, we drove southward to Cape Girardeau County for the night. Immediately after breakfast the next morning X.Y. went collecting, loaded with camera and a large green vasculum made to order. This vasculum, covering the whole torso above the belt when slung over his shoulder, gave the impression of a very wide-shouldered, slim-hipped man gradually disappearing in the distance.

It was my duty to break camp and change the driers. Then if any time was left, I was free to browse around the campsite.

In the first few days of a collecting trip, press changing was a short and simple procedure. Then day by day as presses increased in size and numbers it became a problem to find enough dry blotters or space large enough to dry the blotters. Removing damp blotters from between specimen sheets was not in itself an arduous task, but if the wind blew, two hands and two feet were not enough to keep the numbered newspaper sheet and the specimen together. A major crime could not compare with the separation of a specimen from its numbered sheet, which rendered the specimen useless.

After all the specimens were weighted down, blotters were spread out in sunny fields or on the roadside. There they lay and soaked up the sunshine, or blew into the road to be run over, or into creeks or ditches, or briar patches, or on mats of poison ivy with always the possibility of a snake in the shade of one. They never failed to excite the curiosity of any one who came along. One woman who saw the gray blotters spread about hurried down the road from her home to inquire about the wreck.

As soon as the sun had dried the blotters I would gather up a few hundred. No need for victrola or radio rhythmic instructions, "Bend down, sister, if you want to be slim." When the reverse process of placing the hot blotters between the numbered sheets was completed and straps tightened on the press, I could ramble to see if I could find a rare specimen or new county records with which to surprise X.Y.

After a hard day of hiking it was always tempting to stay in bed late unless it rained, as it did the next morning at 5:15. The drizzle kindly ceased long enough for bacon, eggs, and coffee over a camp fire.

The car was again loaded with our equipment and newly collected plants to head for Scott County. There we passed Benton and drove southward to an old cemetery. Rare specimens are often preserved in old cemeteries long after they have disappeared from fields and roadsides. Aside from plants, the cemetery proved otherwise interesting. The graves had been decorated with items either cherished by the departed one or considered decorative by the survivors.

On the grave of one adult, from head-to-foot of the mound were: one conch shell, purple-tinted glass base of a lamp, golden china slipper with pink flowers on front, small inverted powder box lid, small soda bottle on its side, inverted stem goblet, base of purple glass, China butter dish lid with a chicken on it, glass insulator, a small pair of opera glasses, and a short iron rod.

Another grave had a Mazda light bulb pointing toward the head piece of artificial flowers in a wooden box with a glass front.

In that vicinity we found a field of fragrant Bastard Toadflax (*Comandra Richardsiana*) of lavender color with the fragrance of Lily-of-the-Valley. Also there were many *Opuntia rafinesquii* and Blue-eyed Mary as well. As we walked through sandy areas, thorns of *Opuntia* cactus and spiny *Smilax* plagued us by tearing clothes and flesh, and piercing our shoes, while *Verbenas,* and Cross-vine (*Bignonia capreolata*) thrilled us. We were rambling, not aimlessly as it might appear to the casual observer, but with eyes open for anything new. From past experience we had learned that a plant might turn up where least expected.

Dry sand in this area can be worse than mud or snow to stall a car. Like a bog, the more one struggles to get out of dry sand, the deeper one sinks. Twice our car was dug out and pulled out of the sand with the help of kind residents. We drove all day through New Madrid County making stops for short sallies at interesting-looking places.

Birds of the day were: solitary sandpiper, red-winged blackbirds, crested flycatcher, white-eyed vireo, yellow-throated vireo, Maryland yellow-throat, Carolina wren, cardinal, and a mockingbird, and lark sparrows lined up on a fence.

We paused to tease a colored hog-nosed snake that spread its head and one-third of its body. It could rear the whole spread part above the ground like a cobra to bluff us, but had less luck than the wasp that tormented while we put dry blotters into press.

After passing through the attractive town of Dexter on the peak of Crowley's Ridge, we went to Beech Grove along a white sandy-bottomed creek of soft water where we camped for the night. Again we were mistaken for fox hunters. Men with a pack of hounds stopped to talk on the predatory nature of the fox—mainly the plundering of chicken houses; and the sport of fox hunting. How X.Y. could compete in a conversation with fox hunters, never having hunted a fox in his life, and at the same time put plants into press and write up his notes, furnished me with side chuckles. The idea kept popping up that he might somewhere put the word "fox" in his notes such as: "angles of stem noticeably hairy; calyx-teeth *foxy* at the tip." At any rate he could throw in a sentence now and then that could keep the men talking on their favorite subject. Finally they moved on with the hounds far in advance baying in unison—all except one white hound that chose to spend the night with us.

We were "ahead of the hounds" the next morning at five thirty, heading southward through Dunklin and New Madrid Counties to Pemiscot County, the most southeastern county in the state. X.Y. was trying to reach a swamp east of Big Lake near Hayti. He found the road under water, then tried to walk across a field where the water had almost receded, but found the woods standing in water deep enough for men to row boats. Since it did not seem likely that he could find plants in the woods by boat, he returned and drove back to a swamp south of Deering where he found the beautiful blue *Clematis crispa,* also called Blue Jasmine. It is found in swampy regions of southeast Missouri only in low open ground or trailing over fence rows. This one is more showy than the other related Leather Flowers.

After walking through muddy swamps all day, the swaying strands of *Cabomba* and *Ceratophyllum* in a deep ditch under the bridge where I stood almost set me swaying in concord.

Plant study and bird study go together, for how can one be outdoors all day and night and not be aware of the gay colors and animated spring songs of birds. An indigo bunting chasing a bronzed grackle back and forth across the road could mean only one thing, the bunting's nest nearby. Gray gnatcatchers were busy cleaning the air; redstarts, black-throated green warblers, yellow-throated warblers, prothonotary warblers, and black-and-white warblers were cleaning the trees of the insects threatening the new leaves. Rose-breasted grosbeaks, cuckoo, wood thrush and orchard oriole made their contributions to the balance of nature by eating insects.

We were troubled with muddy roads and rising creeks in Howell County. One rising creek was negotiated, but the second one was too swollen to ford. X.Y. took no chances after he waded part way across and found it too deep and too swift. We had to turn back over muddy roads to Moody near the Arkansas line to make camp. At one time we were stuck in the mud for an hour.

The following day while hunting for the nests of the red-winged blackbird and marsh wren, I saw a sora rail fly from a thicket of Cress, dock and grasses. It flew to the edge of a wheat field and dropped on one side as if dead. I went across and nearly walked over it when it suddenly flew from under my feet, very much alive. He probably suspected I would do just what I did.

Near Tecumseh in Ozark County we camped on the north fork of White River. While X.Y. hunted for plants, I happily made camp in one of the most pleasant campsites. After a steak broiled over coals, boiled potatoes, a salad of cress, and fruit and cake for desert, we went to sleep to the tune of the river "cello," spring branch "violin," tree toad "oboe," owl "bass horn,'' with a special accompaniment from chuck-will's-widow. The starlit ceiling with silvered leaves spread across it, showed not a sign of rain until four in the morning, when an electric storm rumbled up, threatening a downpour.

Since the deep river cut us off from the hills on one side, and a shallow branch came between us and the hills on the other side, we hastily moved across the creek for safe access across to higher ground before the storm broke and sent the little creek roaring into the big river. Sometimes very foreboding threats make only a big noise in passing. Hoping this one would pass us by we made three trips across the creek in bare feet to move our beds without packing them in the car. We needed no flashlight since lightning lit up the creek and river, and thunder rolled from hill to hill. Our trouble was for naught, as we barely got across and tossed the things into the car before the downpour came.

Later that day we walked over some of Missouri's most charming country. In Big Creek Hollow we came upon gorges, waterfalls, and pools of crystal clear water. While skirting along the foot of the bluffs we saw thousands of *Tradescantia ozarkana* in all colors except yellow. *T. ozarkana* is found in only four counties in Missouri, all bordering the Arkanasas line. In nature they hybridize freely with any other species or variety of *Tradescantia* forming a great complexity of colors and sizes. Various species have been hand-crossed easily in my garden. A double-flowered white *Tradescantia ozarkana* produced new plants from seeds of the double flowers. Also new plants were obtained from three other species which were crossed with the individual double flowers of *Tradescantia ozarkana*; however the experiment had to be abandoned before the new plants flowered.

To see one of the richest plant treasures in Missouri, *Penstemon Cobaea* var. *purpureus,* also known from only five counties bordering the Arkansas line in the southwest part of the state, it was necessary to climb an almost perpendicular face of a bluff to avoid a long-distance approach to the limestone glade where the plants grow. The deep rose-purple, inch-long flowers, arranged in loose sprays along a six to twelve inch length of stem, would not persist long in a place easily accessible. Spread over the sunny glade, this exquisite and elusive sight came into view as I pulled up over the rim of the bluff.

The thrill of this exotic *Penstemon* was a fitting climax to end the expedition of ten days. One more stop was made near Montauk State Park, where X.Y. hunted for Honeysuckle by flashlight. The next morning we were up before daylight to start the 500 miles back to Chicago.

Chapter III

APRIL

Visit to a Bog and a Holly Grove

After a long winter in the Chicago area, sometimes before the snow is completely melted from our drive, a few early Snowdrops, Hepatics, or Ozark Witch Hazel blossoms stir the imagination and conjectures on the progress of spring in Missouri, usually two weeks ahead of us. An early spring in Missouri could mean more new county records to add to the "Flora." X.Y.'s aim was nothing short of an exhaustive survey of every nook and corner in every county at each season of the year.

Leaving with a heavily packed car we drove across Illinois noting the advance of spring as we went southwest. On this particular April 15, about 15 miles south of St. Louis we found apple trees and Sassafras in flower. One farmer told us that Sassafras does not bloom. Obviously he had not seen, on the bushes packed tightly along his fence rows, the mass effect of the yellow Sassafras flowers against the blue sky. He would know more about the roots used for tea to thin the blood in spring, or the durability of the wood for fence posts or rails. Now, beyond the present flowers in view, tantalizing glimpses appeared to me from the distant past, and I recaptured for a few minutes the serenity and unmarred security of my childhood through magic blue egg-shaped fruits in coral saucers of Sassafras. Again, the farmer never had made a proper connection between the yellow flowers of spring and the blue and coral fruits on the same bushes in fall.

Farther south along Royal Gorge we found *Senecio plattensis,* False Garlic, Bird's-foot Violets, Red Maple in fruit, and Sugar Maple in flower. Many other of the earliest spring flowers greeted us.

Early the next morning we left for the Bogbean area. Many years preceding, in 1936, X.Y. had found the Bogbean (*Menyanthes trifoliata* var. *minor*) at this place. He revisited the site in 1938 and 1952. And now he was here again. Reynolds County is the only known locality in Missouri where the plant is found up to this date. Though it is known from Labrador to Alaska, in Virginia, Maryland, Nebraska, Wyoming, and other states nearby Missouri, it somehow was left stranded behind in just this one swampy place in Missouri as the ice sheet retreated northward. Now after thousands of years it remains the most southern station in the United States along with several other relict species: *Campanula aparanoides, Liparis Loeselii, Filipendula rubra,* and a few others. The Bogbean is considered poisonous.

As we aproached the area I could see that X.Y.'s spirits sank to depths murkier and lower than the bog. His wrath burst out against the avidity and lack of vision of people in general and of Bogbean owners in particular. The blow hit harder as we came nearer. What had been going on was quite evident. The bog had been cleared, ditches dug, and a disk was left mired in the stiff mud after it had disked the whole area to bare blackness. They were trying to drain the bog and make a cultivated field.

I was hoping his fury, poured out on my understanding ears, would be cooled before we reached the house. The owner of the farm, Mr. G., was at home and he and X.Y. went down across the bare field, which was once the luscious bog, to try to locate a remaining plant. At the very lowest end they did find some. Then followed a lesson in conservation of rarities. Residents, however unlearned in native plants, usually show interest and a desire to cooperate once they learn what unusual plants exist on their land.

A year to two later we learned that the Reynold's County bog had won its battle over man, since all his efforts to tame it had ended in failure. He eventually recognized that a bog remained best as a bog.

Feeling somewhat better as a result of having relocated a few Bogbean plants, we left the farm and went onward to have lunch on the Meramec River bluffs at the crossing on Highway 19 over a tunnel. Not many people who cross the river at this place know the tunnel exists. Though not too long ago, in 1865, an English engineer by the name of Thompson, with a compass, and laborers using only picks and buckets started the tunnel 400 feet through solid limestone rock 60 feet below the ground. Two crews of men started on opposite sides working toward the middle. A small jog in the tunnel indicates a slight error in judgment where the two forces met. The purpose of the tunnel had been the conveyance of water from a spring to a mill to grind grain. Water no longer goes through the tunnel to turn the mill, which has long since been abandoned.

Leaving the tunnel area where the highway cut through the hills, we found *Silene Wherryi*. This plant with rosy-pink flowers, 50 or more rising a few inches above a rosette of leaves, is very conspicuous against a bare clay bank and very charming against a bed of white chert on a hillside. Though one of the most beautiful wildflowers in Missouri it is restricted to a cluster of eight adjacent counties in the central Ozarks.

From Sligo on Highway 19 we left the car on top of the hill and started to walk down to our Ozark farm to make the first spring survey of the flowers there. We aimed to come out in *Crawford County* on the road by the barn *west* of the house; however we made a wrong turn and got lost. The deer we saw in an oat field was not lost according to the calm manner in which he ate the oats, but no such field existed on our place to my knowledge. Finally, by following down a creek which I felt certain would eventually flow into Crooked Creek, we came to the east boundary of our farm at the most distant end on the *east* in *Dent County,* a lucky break. This placed us right in the midst of the bottomland we wanted to explore. We were pleased to find Indian Paint Brush along two spring branches, Dog-tooth Violets, Redbud trees in bloom, Dutchman's Breeches, Bluets, Bird's-foot Violets, golden Bellwort, *Senecio,* five kinds of violets and many others we had hoped to find. Had we been permanently lost here we could have eaten a salad of mildly tart Redbud flowers and could have found enough young fruits to cook like green beans. From there we drove north terminating a weekend trip. Many more April explorations must follow in other counties.

Years later, on April 23, with hopes of adding more records to the "Flora," and more and more crosses on county maps, we left our home in Illinois at one thirty in the afternoon arriving in Missouri at Cape Girardeau. From there we drove westward to Stoddard County near Essex, which was near the home of Mr. W., our chief aim on this trip.

The following morning, after a pancake breakfast with fresh eggs purchased from Mr. W., we walked in this area rich in plant life where Crowley's Ridge joins the flood plain. There is evidence that the ridge lay east of the Mississippi River until the channel of the river was changed by one of the most violent earthquake sequences, December 16, 1811 to February 1812. The flora of this area lends support to the geological evidence that the quake switched the area from Illinois to Missouri.

This ridge is an elevated strip of white sand and gravelly land now containing some "blowouts" resembling the dunes at the south end of Lake Michigan. Some trees in the dunes seemed not too stable in the sand. We investigated the unusual flora on the dunes then went to the spring which flowed from the base of the hills. Along the spring branch we took pictures of hundreds of Yellow Fringed Orchids (*Habenaria ciliaris*) which were coming up in an Alder thicket along the swampy branch. This Yellow Fringed Orchid can compete with the most exotic tropical orchids in color and design. There we found six kinds of Ferns including Royal, Cinnamon, Bracken, Marsh, Sensitive, and one undetermined. The rare Chokeberry (*Pyrus mellanocarpa*) was in flower. This is the western limits of its range, and the only county in the state where it is found. The Chokeberry has much pectin. In areas where found it has been used to make an attractive reddish-purple jelly.

Along the branch also was *Sphagnum* moss, and above all, lovely stands of American Holly (*Ilex opaca*)—Christmas Holly. Tall trees and thickets of smaller trees of this rare holly shaded the banks all along the creek. While it is found from Florida to Texas, north to Massachusetts, New

York, Pennsylvania, West Virginia, Ohio, Kentucky, Illinois, and Oklahoma, it is found only in seven of the most southeastern counties in the boot heel of Missouri.

We took a picture of one tree with a trunk of one foot in diameter and a height of 40 feet. Leaves and red fruits are used for Christmas decorations. The wood has a greater density than Flowering Dogwood. It finishes well and is used for veneer of furniture, musical instruments, black keys for pianos, knife handles, and articles that require a fine, hard finish.

Where the spring branch spread out over the meadow, we jumped from one grassy tussock to another in search of plants peculiar to such swamps. When that survey was finished we returned for a review of the rare thicket of holly trees.

Though this area was so rich in flora, Mr. W. and his neighbor, Mr. M. across the creek, were having difficulty in making the land pay. Mr. M. left his team and plow to visit us. He said that for three years the drought had ruined them and now the fourth year was beginning dry. The freeze had killed all early crops of oats and garden, so they would not make anything this year either. Mr. W. was interested in all possible ways of making money from the farm and was having no luck.

Both men were trying to dry up the swamp by cutting and burning, pasturing, and digging drainage ditches in one of the rarest spots in the state in hopes of coaxing more crops from the soil. The quality of the soil has a great effect on the quality and quantity of plants that grow there. We are just now barely beginning to understand the symbiotic associations that exist between plants. Over millions of years plants have built up this cooperative association which man can upset and be unable to replace in a few years. Here again it is best to let a swamp be a swamp.

X.Y. reflected, "Maybe the soil will not make you a living here, but what a gem to own for preservation, or to save as a preserve for posterity. Here in one parcel exists a spring branch flowing from sandy hills, between banks of yellow-green Pincushion Moss, then spreading out into Alder swamps carpeted with Yellow Fringed Orchids among ferns, all shaded by rare Chokeberry which in turn is shaded by American Holly trees. Each one is rare enough in its own right to justify preservation. I wish I owned it. How much would you sell it for, Mr. W.?" But Mr. W. was not then ready to sell it.

To us the destruction of such rare places and plants is never justified by a few dollars for temporary use. The signature these men were trying to write on the face of this relic was not one to bring renown to them or their descendants. Perception for aesthetic values is neither bought nor learned in later life. To some extent it seems to be inherent. To a greater extent it is built into a young life with frequent exposures to appreciation for aesthetic values that become a permanent and untaxable wealth of a lifetime.

We always left such areas reluctantly, not knowing if partial or total destruction would greet us on the next visit. Our sympathy went to the people who owned unprofitable jewels. This was a case of relative values, where our long term interests for a state or nation as a whole lie in the fact that the irreplaceable jewel was of more consequence than the nourishment for the body where such nourishment could be procured by other means.

Acres of farmland can produce corn to feed animal life, but such land is totally unable to grow orchids to feed the mind and soul. Leave each bit of land to its own. The plants of this area are products of trial and error over millions of years. A stable relationship, unknown to man, has been built up between these plants promoting their own best interests. Man can take away one element of the related whole and upset the balance.

We followed the Chokeberry-lined creek to the swampy lower end of the farm, but when the survey was finished, again we were drawn back to the holly trees.

We spent most of the day exploring this farm. X.Y. did recommend it as a preserve and some time later, the 80 acres were bought by the Missouri Garden Club and Nature Conservancy through contributions to that fund.

After leaving there we followed Highway 60 from Poplar Bluff where we relaxed some from the strain of seeing men trying to combat nature with so little reward for themselves and so much destruction of irreplaceable nature.

Along the winding highway, a sign "35 miles an hour—go slow" would slow us down to botanizing speed. After many windings a sign would appear, "Resume speed—Thank you." But before we could read this sign aloud and get adjusted to 50 miles an hour, another sign would loom up, "30 miles an hour—go slow." That went on for 45 miles of the constantly winding road.

All through southern Missouri we found that the late frost had killed the early foliage of trees, and spring continued to be later as we traveled westward.

Mockingbirds, with a quality of song that only a southern mockingbird can have, added enchantment to the spring that was bursting out in the Ozarks, especially now that rains had come here. A man at Mountain View filling station in Howell County said they have had more rain so far this spring than in the past three springs put together. We reached the U.S. National Forest Preserve in Douglas County for camp Monday night on Hungry Creek, a clear permanent stream about 20 feet across. Except for a whip-poor-will (one persistent whip-poor-will), no-seeums (insects by the hundreds), and an abandoned house (26th abandoned house counted that day), camp that night was very much isolated. Not one person or animal appeared all night or morning except the birds which were unusually plentiful.

At nine o'clock we packed everything into the car and left to hike over the hills. X.Y. and I were soon separated as I could not follow his moves fast enough. I crawled up over a big hill of loose chert, then down to an east ravine. There I followed the rocky branch and lower slopes up to where it branched into three smaller ravines. Not knowing which one of the three X.Y. had taken I waited there until I thought I heard him returning, then I crawled up another hill, over it, and down to a ravine going northward. Since X.Y. never did show up, I followed our original ravine on down to the creek. The many hours I hiked alone, and waited alone, unable or unwilling just to keep up, gave me leisure time to see much I might have missed, and to sense the inspiration of places. How does one identify with, or conjure up an affinity with a local, when one's eyes look only at the path; one's ears hear only the leader's call; and one's only thoughts are hurry, hurry, keep up. With leisure and solitude, miles of physical expanses of hills and hollows expand to infinity. There is no such thing as being alone with nature.

I walked up the creek until eleven thirty when I thought X.Y. would be returning. After all the ups and downs and overs I never did catch sight of him, however the woods were covered with orange and yellow Honeysuckle, and Saxifrage, and Bluets clung to the bluffs and slopes. *Silene virginica* appeared frequently as red splashes against chert, and dogwood trees (*Cornus florida*) dotted all the slopes. In places the apparent dots were really drifts of many trees.

At noon I returned to the car after discovering an old mill site near the abandoned, tumbledown house. Two kinds of rose bushes and a lilac bush under which grew a double-flowered, Orange Day Lily persisted without care. Old-fashioned roses, lilacs, and day lilies are commonly found around old, abandoned houses. Though the house has fallen apart and tumbled down, these plants, once cherished and cared for, persist and perpetuate their beauty and fragrance to delight any one who might come upon them in tall weeds or in a closely-cropped cow pasture. Rows of dark purple and various shades of rose-colored Spiderworts forming a beautiful display on the border of the lawn led me to believe that somewhere not too far away they could be found in the wild, though I had not seen one in all my morning scrambles over hills and ravines.

I returned to the car to get the shovel with which to dig up some roses, double Orange Day Lilies, and Spiderworts to bring home. X.Y. still had not returned when I reached the car with the load of plants, so I walked back to the old mill, now fallen apart. Only the floor of the mill was left. The mill and wheel had been built into a space dug out of the banks some eight to ten feet above the present stream. I wondered how water was supplied to the big wheel until I found that an old millrace had been built between the creek and the hill. The race, about eight feet wide, had filled up with dirt until it now appeared as a depression of an old road between the hill and the path over a sandstone wall on the lower side of the race. It was now covered with Bluets, Buttercups, purple Phlox, scarlet Silene, and mosses in such profusion as to almost mask the identity of the wall. Dogwood arched over the old race with no one to disturb its flowers. It was past noon, yet X.Y. had not returned.

I followed a path from the house down to a spring coming from one side of the creek. The spring water bubbled up with considerable force to join the waters of the creek. I sat down on one of the large flat rocks by the edge of the spring and listened to the music of water hurrying over rapids in the creek, harmonies that constantly changed though the same volume of water seemed to go over the same rocks. There was not a discordant sound. Large boulders had fallen from bluffs above, forming islands of various sizes in the creek. Trees arched over the stream making patterns of shade and sun on the water.

All was conducive to reflections and meditations on the currents in human life that had throbbed and ebbed with the passing of the stream. Ghosts of the past appeared, then instantly followed the current tripping over the rocky bed: a farmer hot and dusty from the fields hurried down to cool his throat and face in fresh spring water then sat on one of the flat rocks to rest; a tired housewife came down for a bucket of water, the only source of running water for the house a block away, and was tempted to sit and rest in this quiet and cool place and forget for a moment the pressing problems of the hot house; happy children ran up and down the creek splashing water to their heads; a woman broken by grief tried to find solace in the tranquil isolation of the spring that lost its waters and yet continued to give; a man perturbed by discordant problems tried to draw a moment of peace from the harmony of the stream; two lovers sat in the moonlight and felt that the tinkling melody of rushing water reflected their joys; another two came to end an engagement and heard only discord in the rushing water; a chastized child came to wash his tear-stained face and felt companionship with the stream. Those and a thousand more involved in the history of the spring paraded down the stream as I sat and mused. It was truly an enchanting place. I wondered how many problems had been solved, how many tears shed, and how many joyous times had been spent here. Obviously no one had lived there for years. I lost all track of time while reminiscing, until hunger reminded me that the afternoon was slipping away.

At two in the afternoon X.Y. returned saying that I had missed the best part of the trip. He had seen thousands of spiderworts (*Tradescantia Ernestiana*) of every shade of blue and rose. My suspicions when I dug the plants at the old house were now verified.

We ate lunch at two o'clock on the rocks by the spring, then immediately left for Wright County.

For sometime I had been trying to find a theme for a talk on conservation. While sitting in the car in Wright County waiting for X.Y. to return from a hike, I decided on the concrete phase of the talk using our exploration of this county as the visual evidence of conservation or lack of conservation, supported by Kodachrome slides I could take while on this trip. It seemed possible to portray three or four generations of housing, farm procedures, and depletion of the land. All could be supplemented by abstract examples from other sources. While waiting in the car for X.Y., I outlined plans for the talk.

In mid-afternoon we arrived at a small country cemetery by a cow pasture. Loaded with collecting equipment, cameras, maps, and notebooks we began to hike down a steep, rocky hill holding on to a strong four-strand barbed wire fence. We were one and one-half miles southwest of Cedar Gap on the headwaters of Bryant's Creek. X.Y. carried the maps as usual. After leaving the fence and crossing a large ravine we climbed a steep, bare hill to a pasture where black cattle had grazed acres and acres to the bare rocks. Not one green thing was visible on the ground. Cattle were standing on their hind legs to reach the branches of trees that still had a few leaves left.

We took pictures of the large hill of bare rocky ground on one side of the fence and the green grass, lush plants and shrubs on the unpastured side. On the pastured side there was not a discernible piece of soil on the whole hill. There was nothing there to stop the speed of water that fell as rain or to hold it long enough to soak into the soil below the rocks. It seemed to me I walked a mile on those bare slopes, so steep I had to dig in to stand up. The sun was hot, and we descended, crawled through the fence, and went down a branch on the ungrazed side.

From there we hiked down very interesting slopes and branches until six o'clock. The creek had a habit of suddenly disappearing for a few hundred yards, then coming up again from the dry bed in a strong stream. X.Y. was getting ahead. Wood thrushes were singing. I sat down on a large rock in a part of the dry creek bed to relax and lose myself in the song of the wood thrush. Each clear note vibrating with overtones seemed to fill the dense forest of the bottomland already growing dusky. Hundreds of evenings this song goes on unheard except by the denizens of the forest. This evening I was there to exult in what would ordinarily seem to be a waste of a sublime serenade in an inviolable setting.

At six o'clock I pulled myself from the spell of the place and turned back toward the car. X.Y. had already gone on too far to hear my "Yohoo." I got along well, I thought, until I reached a fork in the large ravine above where we had entered the main ravine from the pasture. I was trying to go back an easier way than crossing the bare, rocky pasture. Moreover, I did not like the looks of one or two members of the pasture. I had no idea which of the forks of the ravine to take. The north fork seemed too far north, so I took the south one. The sun was already down behind the hill and out of sight. After a time of climbing I came to a fence which I followed up to a glade where the fence ended. Now, like leaf-cutting ants that have reached a disturbed part of their trail, confusion set in. I went back down to the main ravine to another fork which I followed up to the top of the hill. I felt that I was too far north of the location of the car. Also I thought that if I kept on until I reached the top of the hill, I should come to a road, though I would not then know whether to go north or south on the road. While the ant uses a sense of smell to follow a trail laid down by a scout, or a formic acid trail laid down as a roadway, I was trying to use the sense of sight that often fails on a back-track: nothing looks the same from the other side.

About the time I was beginning to feel a bit panicky, I saw a house in the woods at the top of the hill. I approached and asked a small boy on the lawn if his mother were home. The grandfather came from the house. I said I was lost.

"Where abouts have you been?" he asked.

"Bryant Creek Hollow," I answered. "I'm sorry, I got confused."

He then said, "You don't have to apologize about getting lost down there. I've lived here all my life, and I get lost every time I go down there. My brother, who was raised here, came back from California last summer, went squirrel hunting down there and got lost. We had to hunt for him."

I felt better. Then I said, "We left our car parked by a cemetery. Am I north or south of the cemetery?"

"You're two miles south of the cemetery by ridge road, only one mile across the hills, but with darkness coming on you'd get lost again."

"I'll take the road," I said feeling sure I'd get there, but not ahead of X.Y., unless he got lost too.

The boy added, "Yes, and there's a hole, a big hole in the woods. You might fall into it in the dark, if you go through the woods."

By this time the mother, carrying a baby in her arms, came from the house. The man said, "If you don't mind riding in our old jalopy, we could take you over to the cemetery."

The mother then told me that an eight-year-old girl who had eaten the root of a Snakeroot plant on her way home from school had died. Snakeroot is a confusing word, a common name represented in six different families of plants. Some have been used for medicine and some are poisonous. White Snakeroot poison has even been transmitted to people through milk from cows that have eaten it. In this case the root had confused the girl who had mistaken it for an edible root. They said she turned black in spots. She had been buried in that cemetery the day before. The woman said that she wanted to go to see the flowers on the grave so she could make some artificial ones like them.

"I'll be happy to ride along, if it's not too much trouble," I agreed, trying to hide my joy at getting there so fast. I was tired, however it was not the two-mile walk along the ridge road that bothered me. Comradeship with, rather than fear of the dark woods posed no problem. But the pandemonium that would ensue if X.Y. arrived at the car before I did was a fear approaching panic. He might complicate things by starting for help, thinking I was lost in the woods.

The old Ford rattled up to the cemetery at 7:15 p.m. The family and I clambered out just 15 minutes before X.Y. arrived. I never did tell him I got lost.

We made a wild dash for a place to camp. By the time we found a suitable place and the potatoes were cooked and the tuna salad made, it was ten o'clock.

I had made a good beginning on my conservation lecture, and X.Y. had added more records for his book.

A cardinal in the tree above me the next morning kept singing the wrong song by singing one syllable when there should have been four or six; a young bird, of course, just learning to sing.

Judging from the name Beaver Creek in this county, beavers once helped in flood control. Now the creek which drained the surrounding bare, chert hills, was bone dry even in spring. Also the name of Elk Creek hinted of a lusher time. Here, no one had realized that the people who use the watershed hold in their hands the lives and well-being of the millions of plants and animals on it from its origin all along the way to the sea. Their foresight did not extend that far.

The capacity of ground to hold water back depends on the quality and depth of the soil and its cover. Here, over much of Wright County the cover is gone and with it the soil. A good soil can hold more than one-third of its bulk in water, and here, there was no soil to hold it.

It takes some 500 years to make an inch of soil and less than 500 days of heavy rains, overgrazing, poor soil management, and fire to destroy it. Here we saw the evidence of burning, girdling of trees, and the shift from too many wheat and corn crops to cattle and finally goat grazing. Homes ran from the log cabin and stone chimneys of pioneer days to the eight room, two story homes of prosperous farming and cattle raising days around 75 years ago. Some were now gray from want of paint, and others abandoned. Modern homes were now only three or four rooms surrounded by bare, chert hills scarcely able to support a goat. Here was my whole paper in one parcel.

At a filling station I asked a man what he thought about so many goats. He answered, "Just about everybody is going to goats. People are trying to get rid of *bresh* to grow grass for grazing."

Then when I asked about killing the trees he replied, "Too many trees now."

What he could not see at a glance was that when the soil was gone neither he nor his children would ever see anything grow on those hills. Neither could a botanist ever collect any flowers from them. In three to four generations they had drained the land, the last ounce from the land.

The small modern houses bore television antennae. Also near the top of a hill, on a one-way rocky road barely negoiated at ten miles an hour was a sign, "Radar Controlled," a humorous quip carried out by some local talent.

I wondered, could the author of that sign have been inspired by a vehicle similar to the one we had just met coming down that road. On a chair in the back of that farm wagon speeding to a church dinner sat a woman. As the wagon jolted over the rocky road, her problem was to stay with the vagrant chair and manage the cake on her lap so as to compensate for the inertia of sudden moves.

A day later, driving northward, we stopped at Moniteau Creek. There we walked over north-facing bluffs covered with Fragile Ferns, *Isopyrum, Senecio,* dark purple Larkspur, Spring Beauties, *Anemonella,* Wild Hyacinths, Yellow Star Grass, and *Hydrophyllum* in solid masses over the hills. Nature had not been disturbed there. Leaf mold was thick enough to catch and hold the rains for plant use, prevention of soil waste, and reduction of floods.

At camp, the no-seeums were there, but not as bad as at the preceding camp. X.Y. wore a towel around his head looking like a Turk. Here we saw a flock of goldfinches eating ripe dandelion seeds.

Along the river bottoms where the ground was blue with Bluebells (*Mertensia virginica*) a man wanted to know if we were fishing. It always seemed difficult for anyone to understand what we were doing in the woods or along a river unless we were hunting or fishing. Then when told we were looking for plants, the next guess was plants for commercial or medicinal purposes. No one-sentence answer could clear up their questions.

At one the next morning I was awakened by rolling thunder and lightning. We barely had time to scramble everything into the car before it began to pour. It is nice to be dry at that time of the morning, even if there is no place to put your feet and no pillow for your head.

At Boonville later that day we stopped for groceries. Farther north at Richmond, we found spring far in advance of any place we had been in the south. The late frost had not killed back the northern plants since they were not so far advanced in the north. However near McFall, in Davies County where we camped, the temperature went down to 40 degrees, and people expected frost. Whip-poor-wills called all night, and a barred owl was hooting at a slower tempo. We made a camp fire to keep warm the next morning.

While eating our breakfast of grapefruit, scrambled eggs with cheese, and toast and coffee, a man came by and asked X.Y. if he were fishing, then quickly said, "No, I see. You're the plant man. You won't find much on these hills. There's common dirt in the bottoms, but there ain't much on the hills."

He remembered X.Y. from the July before.

Near Trenton very early, even before sunup, we hiked down many ravines. Near a rocky spring branch I found a Showy Orchid (*Orchis spectabilis*) by an Indian Paint Brush. X.Y. had stepped right over a clump of orchids in bud without seeing them—the only orchids on the whole trip since the holly swamp.

With many stops to check for wildflower records, we drove eastward to Clark County. There X.Y. found *Oakesia* and *Arenaria lateriflora,* the latter being the third county record in Missouri at that date, and that was found accidentally.

Sunday we left Missouri by way of Keokuk, Iowa, for our home in Illinois having covered 1,731 miles in one week.

CHAPTER IV

APRIL

Farewell to White River

April is a month of hopes—hopes of greeting flowers and future fruits, hopes renewed each year with fresh zest in spite of repeated past failures, in spite of untimely frosts, not enough rain, too much rain, the wrong soil, insects or disease, any or all of which cut short perfection in the past. Yet, again with each new year, spring signs combine to brew a new courage within the soul, whispering that this new spring will be different, the one always hoped for.

However our mission on this 24th day of April was neither one of hope nor expectancy, but one of farewell. Not farewell to one season alone, good or bad, as the case may be, but to all seasons that had passed over this particular section since the beginning of time. We had come to bid farewell to a river and all that grew on its banks and bluffs, before the whole scene slipped from existence beneath the load of murky water. We had come to glean any possible information about the plant life that existed in the area and to preserve it by dried specimens and by records in the "Flora" under preparation.

White River on the southern slope of the Ozarks drains an area of over 6,000 square miles of forested hills and fertile valleys. White River had been one of the most beautiful in the Ozarks, perhaps the favorite river for float fishing. Float trips of one day or of weeks carried city-tired men and their camping equipment through clean, clear water, over rapids, past gravel bars, around bends to quiet pools, dark green and cool under high rugged bluffs, then back out again to sunlit or moonlit current, deep whirling pools, and on down past ever-changing scenes—with a bass now and then pulled from his retreat under a submerged boulder. But White River with its present and proposed dams was doomed to go. The nearly completed Bull Shoals Dam would back the water up White River and tributaries to a distance of almost 90 miles in Missouri, covering up forever the existing plant life. Who in the future would ever guess that below the water flower-covered bluffs more picturesque than can be conceived in the abstract once existed. Movies and colored slides should have been made to record what once could be seen along the length of that river soon to become a lake for turtles and fish.

We had come to salvage by records and, in rare cases, by transplanting to safer places as much as possible of the rarer plant life. The animal life we left to take care of itself or perish in the attempt to adapt to a new environment. On this day we hiked down the Little North Fork of White River near the Arkansas line. X.Y. made surveys of existing plant life on each side of the river before the water from the soon to be Bull Shoals Dam would come up to bury all existing flora. The day was hot, and I cooled my feet by wading along the solid rock bottom of the river. It was easy walking 50 feet or more of level bottom then take a step down, level spaces and more steps down—all solid rock bottom.

Having little more to do than keep cool, I practiced the game of "before and after, now and later." With closed eyes, the cool water flowing over and beneath my feet suggested the future of dim translucent light through a depth of surrounding water. At present I could come back by opening my eyes to a view of wooded hills where branches of trees dipped into the river at my right. To the left fields stretched out across the valley to a farmhouse. Repeatedly I could close my eyes and go back under, then again open them and imagine what would not be there a month later.

We were not the only ones there to say farewell. Farmers and their families who had lived a lifetime by the river had come down to catch a few last fish where formerly they had been free to run down to the river after a hard day's work to take a dip; or in cool, contentment—with pole and line—sit on a tree root right over the best pool that nature had ever fashioned for river bank fishermen.

Today whole families had come down for a last try at their luck. The wives were preparing to cook the catch over live coals on a gravel bar. They had little else to do at this time. All their bottomland farms were either lying idle or in crops that could not be harvested, because the government now owned the land and refused to let the crops be harvested. They must take their bit of money, buy a few acres of rocky hilltop above the high water mark, cut the trees off of the rough land that could grow nothing but trees, and let the meager soil wash quickly down to help silt up the coming lake.

I talked with one farmer's wife who was sick at heart. "My home where I have lived all my life will go under in a few months," she said. "It's a good house. All our good farmland and the rich farmland of all our neighbors will be 100 feet under."

So much loss and heartache to satisfy an engineer's dream.

As I looked across the beautiful valley to one large, white frame house, a home for more than one generation, and back to the present scene of the neighbors assembled on the clean gravel bar, it was clear to see that their attempts at gaiety were as mottled with grief as the patterns of sun and shade that fell on the gravel bar and on those who passed back and forth under the tall trees. The young children living only for the present day were enjoying it to the fullest. Time would quickly wipe out the memories of their valley homes. Some of the larger children helped their parents set up the improvised supper table of a door on saw horses. Checkered cotton table cloths gave a festive air. Bowls of salad previously prepared from their own gardens were pulled from baskets and arranged at intervals on the long table. Homemade breads, homemade butter, pickles, and cakes made with their own homemade butter all took their places for the last time at this banquet. Artificial butter was destined to appear on their hilltop tables.

I smelled the smoke of the wood fire being reduced to coals for frying the fish then being cleaned at the edge of the clear stream and longed to be present to smell and taste the hot fish when it appeared on the table, but we would have to hurry onward.

The entire present view as compared with what was to come, passed before my consciousness with ironical implications—all these valley homes and farmland 100 feet under just to reduce the flood stage by three inches over Mississippi homes already 30 feet under water. In this enigma I became one of the present assembly on the gravel bar. I could see that the heart of these people was gone and the spirit with it, by the way they moved.

It was no mere collection of words or hypocritical consolatory sentence when I said, "I know how you feel."

One woman asked me, "would you like to see the big spring yonder at the foot of the bluff across our meadow and have a last drink? It is a pretty spring."

I thought of the hot walk, one-fourth of a mile across the sunny meadow, but accepted the invitation. A cool drink was tempting, besides I did want to see a spring that never could be seen in the future. It was too pretty to cover.

As we started, she pointed out a giant tree on the gravel bar near the river. "That big Sycamore standing there 100 years, all of those big trees down there will go under. They won't let us touch them. There's enough timber right there to build us a new house on the hill. What good will they do on the bottom of the lake. No good comes of wasting anything," she lamented.

Their lifetime concepts of conservative living, now brushed aside by complexities not well understood, were changing their past problems of survival to a present problem of rehabilitation.

I was thinking also of all the birds that had nested in the Sycamore branches having to move to the hills next spring, contrary to their urge to return to the old site. They too would feel defeat as well as all other forms of animal life that would have to climb to the rocky hilltops to hunt new homes. "You're right," I agreed, thinking especially of how much I would give for a dining tabletop of Sycamore wood and of how many tabletops such a tree would produce if it were solid. I too was becoming afflicted with conservationitis. Such a tabletop, now beyond my means, as well as a whole log-length of such tabletops, would gradually succumb to bacteria and its fine grain would fill up with silt, becoming soggier and soggier, sinking lower and lower into the lake.

The Sycamore is the largest deciduous tree in the United States, that is the largest in diameter, though not the tallest. The largest living one with a circumference of 43 feet and three inches at five feet above the ground is located near Worthington, Indiana. The Sycamore sets a lush table for chickadees, juncos, canaries, and other finches.

That cry of utter waste came often from people who all their lives had had to consider economy.

When we returned from the spring, X.Y. had completed that small proportion of the survey. With an accumulated load of injustices that would live far into the future, I picked up the bundle of Eglantine (Sweetbrier) rose bushes dug from the gravel bar. The Eglantine rose, a species of rose native to Europe, was well introduced in many places in North America. The single pink flower is not the whole of its charm. The finely cut leaves have a delightfully refreshing fragrance. It is very hardy and able to survive in almost any habitat. On our rambles in Missouri we found it on steep rocky hillsides in full sun, on flooded gravel bars in shade, and on rich soil along old fence rows. It also grows in the heart of those who encounter it often.

Then with one last look down the river, we left, going farther up stream, leaving with sad memories and one Sweetbrier rose bush. I went on ahead following lower slopes along what had once served cows and fishermen hunting for richer pastures. X.Y. followed more slowly trying to see and list the rarer plants as he walked. Among them were: Green Trillium, Harbinger of Spring, many kinds of violets, Shooting Stars, Buttercups, Fire Pinks, Golden Ragwort, Dogbane, Honeysuckle, and others too numerous to list.

Trees had been cut and left in place where they might have come near the surface of the lake. The path, unused for some months, by cows and fishermen, was grown over by weeds and vines. Soon I was aware of a noise in the rear and looked back to see X.Y. face down. With eyes searching the slopes, pencil and paper in hand, thrusting his feet along to find the path, he had been thrown forward by an unexpected vine and landed on a very irregular patch of boulders. He untangled himself from his press and the vines and began rubbing his chest. I retraced my steps to investigate. He had hurt his chest. "Cracked a rib perhaps," he laughed.

I urged him to return to the car and go to a doctor for an examination, but he would have none of that. Regardless of the catch in his chest he went on and was soon out of sight. Nothing short of catastrophe could be permitted to consume the time that might add one more county record to the "Flora."

Near the top of the hill along the river were many kinds of Asters—*Aster patens* with top-shaped head and clasping leaves; *Aster anomalus* with spreading, curling bracts; *Aster turbinellus* with top-shaped head and slender leaves; *Aster linariifolius,* a stiff-leafed one on dry glades—made up a few of the 29 species and 35 varieties and forms in the state.

By now I had given up hopes of following X.Y. and began a search for interesting rocks for our rock garden at home. With the car close at hand I could sneak in more rocks than was possible to carry from distant places. Rocks eroded in interesting shapes, moss-covered rocks, lichen-covered rocks for sunny spots, all piled up under the car seats. We were adding enough queer rocks to our Illinois property to confuse a far-future geologist. A rock has quality and permanency. Its charm endures through years without care long after a flower has faded and lived out its lifetime. Its appeal to creative imagination and meaning has no equal to those few endowed with such perception.

X.Y. soon returned and we drove to Bratten Spring Creek, a mile north of Pontiac where we camped at "Firefly, Gray Horse, Gnat, Curious Man Camp." That balmy night the fireflies across the meadow were as thick as stars on the clearest November night. They were numerous enough to give the impression of twinkling rather than moving about. Gnats were no less numerous. There was only one gray horse, but he stayed all night, and only one man, who moved silently away after satisfying his curiosity about our presence.

The next morning X.Y. hiked while I broke camp, stopping frequently to listen to the songs of mockingbirds and orchard orioles. X.Y. found little unusual there, and we left early, driving to another place on the Little North Fork of the White River north of Theodosia where we scrambled over bluffs spotted with the all-too rare Ozark Spiderwort (*Tradescantia ozarkana*) in every color except yellow. The white, pink, rose, lavender, and various shades of blue petals, delicate as gossamer, are sun shy. In full shade or on cloudy days the flowers persist for a day, otherwise they vanish like frost under warm sunshine.

At noon while eating lunch we sat down on the rocky banks of the stream and dangled our tired feet in its clear water for the last time. The ripple of water over our feet ticked off the time leading up to the end of its freedom to flow. Lists of plants to be drowned grew as the day passed. Blue False Indigo (*Baptisia vespertina*) on the glades, Dogwood, Redbuds, Celandine Poppy, Squirrel Corn, Columbine, Viburnums, rare ferns, and many others on the slopes would be submerged.

After lunch we drove to Swan Creek and on to Swan. Swan, a town of one general store, a post office, and three homes, has character. Though I saw it only once for about an hour, it was the nearest place to an "Isle of Contentment" that I had ever encountered. On the porch of the low rambling store, the owner, Mr. B., sat in an old rocker. Large, wide-spreading trees hovered over the store, the surrounding lawn, and tidy home across the drive in front of the store. At my left a crystal clear stream, Blue Creek, a tributary of Swan Creek, with fall enough to make music, flowed over clean chert, unsoiled by silt from the densely covered hills above. Slanting sunlight through the trees lit sparklers in the rushing stream and formed shadows which danced on the clean, white chert of its banks. I sat in the car and listened to the store keeper tell X.Y. about Mr. B. F. Bush, a former botanist, who had been his guest years ago when he botanized on horseback before the day of cars.

This particular part of the country would be safe from the dam invasion. Mr. B. sat and rocked as his neighbor from down Theodosia way spoke of the dam. "At the beginning we all protested the best we knew how. We tried to organize, me and some other fellows, but they brushed us off like so many beggar ticks. We just have to go to the hills."

I saw the muscles in his face tighten and felt what was going on in his mind.

X.Y. sympathized, "I've done all I can to halt the large dams. There are four viewpoints: First, the home owners like you who grieve to give up a lifetime home in order to lower the water a few inches over some one else's already flooded home somewhere down the Mississippi. They tell me that if all the 19 proposed large dams go in they will lower the flood waters three inches in Mississippi. Second, there is a realtor's point of view with their one big haul for small rocky hilltop sites along the lake for homes and amusement centers—lit with neons and loud with jukeboxes—a place where city people can go to the country and have as much of the city with them as possible. Then third, there is the conservationist's viewpoint which faces the facts that the lake will silt up in one's lifetime; the unsightly mud flats over which nothing can pass across, or along the shore following water fluctuations; the destruction of rare plants, timber, and best bottom farmland; elimination of some of the best springs and clear, clean streams. And fourth, there is the engineer's viewpoint and that is the one that ticks."

"They want the darn dam, and they'll get it," the neighbor said.

"Got it," X.Y. corrected.

The storekeeper rocked on and calmly commented, "I'm sorry plumb to the bottom for you fellows and thank God they didn't build that dam high enough to back the water up this far."

Thus the bodies from buried valleys sadly go to the hills while their souls live in their sunken homes. I was glad Swan was to live on. How it got that name I never learned, yet this small spot reminded me of the swan with its beauty, calm and graceful coordination drifting serenely through life. Like a rare painting before which one stands and loses all sense of time and place, this particular place had infected me with a sense of peace and contentment. I should like each year to revisit the spot briefly to recapture that subtle emanating something that gripped me that day in the peaceful valley of Swan.

As the sun sank we left and went to camp near a bridge over Swan Creek. About four o'clock the next morning a thunder storm came up. X.Y. had to go into the car and try to level a spot where he could curl up in some degree of comfort while I lay snugly on my cot and recalled how he had scolded about my fussing around the evening before. There was reason to believe it would rain before morning. I had carefully selected a raised spot where water would drain in every direction, then covered the mosquito cot tent with the nine foot plastic square I carried under my pillow slip. Let it rain hard; not one drop would come through. Above the splash of the first heavy raindrops I heard a large flock of birds rapidly flying downstream ahead of the storm. During the heaviest part of the rain when my roof began to sag with its pool of water, I would push my finger up against the pool and drain off the load.

About seven o'clock the rain ceased, and I started a fire for breakfast, using the wood I had collected by flashlight and packed under the car the previous evening. Trial after trial it smoked and flickered out. Even newspapers burned poorly. Finally enough twigs dried out sufficiently to blaze. Warblers and wood thrushes sang in the wet woods while we ate our breakfast of fruit, bacon, toast, and coffee. The wood thrush has the rare quality of holding four notes simultaneously. To me it is the most melodic of all bird songs. The first low phrases of the complex melody of modulated musical notes seem to announce the coolness of the damp forest. Then follow phrases which rise in pitch and volume as if amplified and reflected by the wet leaves polished by the rising sun. The reclusive thrush continues from one place in the woods announcing his presence only by song. No other song fits more aptly with the silent, moist woods of early morning or late evening.

It began to rain again before we completed packing to drive back to White River at a point of Horseshoe Bend in Taney County. Since the dam had been approved, the road down to the river had been permitted to wash and degenerate to a point where we had to leave the car on a rocky glade and walk one mile down. Near the base of the hill we struck out through pathless thickets still quite wet from rain. While checking a steep rocky slope for Ozark Spiderwort, Alum Root, Waterleaf, *Corydalis, Senecio, Euphorbia,* and whatever else might be found there, a downpour came, conveniently giving us time to take shelter under an overhanging bluff. Deep-water fish would hide there later. While there, we collected mosses and ferns from shady nooks and bluffs.

As X.Y. zigzagged up and down slopes faster than I could, I worked my way, when the rain ceased, to a lower slope and maintained the level for awhile. When the second heavy rain came X.Y. was near another sheltering bluff while I was stranded below. A large leaning tree showed its bark dry on the lower side, and there I backed up against the dry bark finding the slant sufficient to keep me dry too. Since there seemed to be no danger from lightning in that particular rain, I could take that chance. But why try to keep dry? Every tree, bush, and weed shed its water as I brushed by. Men fishing on the river had given up and were crouching in their boat as the rain soaked through their clothes and splashed upward from the river. They rowed across to the opposite side to their still existing valley homes.

I was reminded of my isolation one mile from the car on the hill, when a movement between my feet revealed a large copperhead snake, newly washed to a clean reddish-brown. He did not like my intrusion either and slid between my nearly paralyzed feet to a bush about eight feet away. I called to X.Y. who came down the hill to dispatch this one—the first poisonous snake I had encountered in years of hiking over the rocky slopes of the Ozarks.

We continued downstream. When the third rain came, I again parked under a conveniently leaning tree while X.Y., already soaked, went on. After fording a few slippery branches flowing into the river, having lost the sound of X.Y.'s voice, I retraced my steps. He was looking for a spring one mile farther down the river, and I would only meet him coming back or miss him should he return on top of the hill. A fourth hard rain came and caught me without a leaning tree. At any rate by now, my head was all that remained dry.

As we had come down the gullied road earlier, X.Y. had pointed out a ravine at our left which he told me to explore in case I turned back before he did. This I went to do leisurely, and when finished, thinking I was still far ahead of him, I started slowly up the hill leaving scout trail signs of sticks and rocks at intervals indicating I had gone up toward the car. When barely a fifth of the way up I heard him calling. The sound appeared to come from down the river. I answered and tarried waiting for him to catch up. The calling continued, but never seemed to move. Finally I gave up and started slowly up the hill toward the car. X.Y. had been there for one-half hour waiting frantically. He had not heard my answers and was about ready to start hunting for me. That was the first time in all the years that he had ever reached the car first and waited for me. In order not to delay his work I always managed to get back first. But what is good for one is good for the other, I thought. He would know how it felt to wait and wonder. Many, many times I had waited for hours, even until after dark, wondering where or how to find him if need be. Sometimes the pictures I conjured up were varied and wild.

On a rain-washed rocky glade we changed to dry clothes and drove to Beaver Creek for the night where the stars twinkled over us until dawn.

After breakfast we drove to Forsyth, touched on one side by Lake Taneycomo, a bulge of the White River, formed by damming for hydroelectric power. At Forsyth we purchased groceries then drove through Branson and Hollister toward Long Creek, where the proposed Table Rock Dam would put a third bulge in White River.

We drove down toward an old ferry on White River; sat in the car and ate lunch in a lane near an old barn not far from the starting point for the afternoon hike. I backed out when it began to pour rain again. X.Y. went on alone across the field, a drenched figure becoming dimmer and dimmer as distance added extra layers of heavy rain between us. There is a pathos at such a picture of devotion to one's drive to carry out a project. Personal comfort, and even health, give away to such compulsion.

I sat in the car watching eight varieties of chickens, two bantams, some pigeons, two young calves, and a pup scramble for shelter. When each had found a place to his liking, I read for awhile, then remembered that the plant driers would have to be changed. There would be no chance to spread the blotters out to dry in the sun. Likewise there had been no sun the day before. Green plants soon darken and mold if not dried quickly. Houdini-like I began in the front seat of the car the seemingly impossible task of removing 200 damp blotters, each 12 by 18 inches, replacing them with dry ones between the folded sheets of newspaper containing the plant specimens. Two hundred damp blotters piled up on my feet as the eye-level stack on my lap diminished and the stack by my side rose above the shoulder. Numb from sitting motionless lest the stack tumble over, I completed the job safely and managed to get the straps around the press. Once a plant specimen gets separated from its numbered sheet it becomes useless. To let a plant press tumble over and scramble the specimens would put me in the doghouse for life.

Late in the afternoon X.Y. returned and we drove to a farm on Long Creek. Mr. S. who had recently moved there from Seattle graciously permitted us to camp on his part of the creek which, to him, meant nothing more than water for his stock. No fortune could tempt us to disturb such a stream and its bluffs, could it be transplanted to our home. Two nights we camped on this "Swishing Creek, Singing Bluff, Rocky Ledge Camp." Isolated from sight and sound of civilization our camp was spread out on clean, level rocks by the edge of crystal clear water running over a solid rock bed. We ate and slept by octaves of music ever varying as water tumbled over a series of ledges and echoed back from the opposite bluffs. Not one mosquito and very little dew disturbed our peace.

The next morning was clear and warm. This spot was soon to be buried by water backed up from Table Rock Dam connecting with Lake Taneycomo on the west and putting a final bulge in the Missouri part of White River.

After putting the camp and myself in order, I waded from one side to the other of the 70 foot wide creek looking up and downstream at its scenic treasures. X.Y. collected at distant points until noon. After lunch we both waded up the creek along the bluffs searching for any rare specimens or information that would be destroyed by the lake.

At four-thirty we returned to camp a second night at the same place. To make camp twice without moving to a new location was a rare occasion, and for that reason we never carried a tent.

Early the next morning we took one last look at this beautiful spot and turned our backs on it with a feeling of finality similar to that one experiences when taking the first step away from the unfilled grave of a dear one. We had done what we could to preserve a few bits of the place, and now all else would be hidden forever by a great depth of murky water.

CHAPTER V

MAY

Orchids, Violets, and Cryptogams

In May, thirty years before our explorations, one botanist had collected a purple Paint Brush (*Castilleja purpurea*) north of Springfield in Greene County. It is also found in the western states, however it had been found only once in Missouri and that in Greene County. X.Y. had hoped to find it again. Also on this trip he was seeking one special Honeysuckle to add to his future "Flora."

Hence at six in the evening after work in the city, we started for Missouri. Driving through St. Louis at midnight, at about three in the morning we landed at Dry Fork where we made camp under starry skies. After nine hours of driving we did not lie awake to look at the stars. The rest received by three hours sleep in the open is far more refreshing than the same number of hours indoors, providing one is comfortable. All cares and problems, in fact the whole past simply drops away like last year's leaves, as if we had no further use for the type of existence from which we had driven only a few hours before. Whether we are out for three days or three weeks, reality is never reestablished until we are almost home.

Regardless of having lost half the night, we were up early, and I followed a bluebird and found a nest with young in a fence post along the stream. X.Y. hurried on over the bluffs while I broke camp and packed the car. I still had time to climb to the top of the bluffs south of the road where I could look out west at a glorious early morning scene of May 5. Mist was rising from the river and valley; dew was still dripping from the budding tree limbs; farmers in the valley below me were making audible contacts with their livestock; birds were singing in joyful celebration of new mates, nests in the making, eggs laid, or young ones hatched. The razor back ridge on which I stood was white with chert outcroppings, and there against the white chert I had my first view of the pink clusters of *Silene Wherryi* in its natural habitat. Among them were the light blue and the bicolor blue-and-purple Bird's-foot Violets (*Viola pedata*), red Columbine, and Yellow Star Grass. No more colorful or charming rock garden could have been planned or maintained by man, yet here it was with no help from anyone.

The Bird's-foot Violets or Hens and Roosters, with pansy-like faces and bird foot leaves, march straight ahead through the years from the age of three to the present time. One little Bird's-foot Violet, the whole foundation in a naturalist's life, grew in a crack of a large erratic boulder below our orchard in Illinois. Because of its sheltered southwest exposure one little greenish bill-like bud made a precautious appearance. It was not to be touched because its face might not develop properly. Each daily visit showed a bit of change, a bit more blue in the swelling bud. One bright morning I ran up through the orchard to the house announcing that the little face had come out. Nothing has the appeal of the Bird's-foot Violet.

Clay, sandy clay, gravel, silt, the most sterile, unproductive-looking soils along the roadsides of the Ozark hills, become transformed to blue swaths of Bird's-foot Violets with the "hens," all pale blue, predominating. Scattered among the hens in fewer numbers the "roosters," with lower petals of the same blue as the hens and the two upper petals of deep blue-purple velvet, arrest the naturalist. On such roads where traffic is no problem, speed can slack or pause to admire as a whole or in part these displays, often backed by white drifts of Flowering Dogwood which seem to dominate the whole leafless forest.

A trick we used to play (anti-wildflower preservation) was to hook a Hen and Rooster by their spurs then pull to see which head came off first—Hen or Rooster.

We left there about ten in the morning for Montauk State Park south of Salem where, on north-facing bluffs above the spring, X.Y. found the Honeysuckle we hoped to find. I collected a specimen of *Mitella* along the bluffs below the spring.

Having found the Honeysuckle we moved on to Percey's Cave six miles north of Springfield. The owner of the cave went with us to help hunt for the purple Paint Brush, which we failed to find. After an unsuccessful search again the next day, we went farther west. This did not mean that the purple Paint Brush is there no longer. A few people searching for a few hours cannot cover every acre of a large territory.

The night on Turnback Creek in Dade County with dead trees rocking in the high winds was not one conducive to rest in preparation for the long drive from southwest Missouri to Chicago. We drove half the night in rain arriving at two-thirty in the morning after 561½ miles of continuous driving—without having found the Paint Brush.

A year later, on May 23, we crossed the Mississippi River at Hannibal. After a camp near Louvre River in Montgomery County at "Hoot Owl, Bullfrog Camp" we hiked all morning along Pinnacle Creek where we found a mother duck with six little ones a few days old. They were nearly black with a buff line over the eye. X.Y. caught two, starting a panic in the duck family. After cuddling the downy ducklings for a few minutes he put them on the ground where they waddled in great haste to join the mother. But we were collecting plants not ducks.

We climbed to the top of the pinnacles to take photographs then back down to the car and found a road where it was possible to drive over the "skyline drive" with deep ravines and canyons on each side. Going downhill another way, X.Y. drove while I directed from the ground as we made our way down the old washed-out, dead end road to Stinson Creek. There we camped in the cold dew of 42 degrees.

We had finished breakfast when a woman and two boys came to get some horses, two gray ones and a bay, from the field near us. One hour later Old Bill, the bay, ambled back, followed by the woman on one of the grays. Old Bill must have decided he liked us during the night or he had become curious about our camp. He came up and let me put a rope around his neck. The woman then led him away and sent the rope back by the boy.

Getting back up that washed-out road was as much luck as engineering. X.Y. had to keep the two wheels of one side of the car on the narrow ridge that stuck up between the two old ruts. Backing up the hill in front of the car, I directed "one-half inch to the left, one-half inch to the right," and so on for what seemed to be one-half mile. Had the wheels slipped off the middle ridge, we would really have been stuck with many miles to walk for help. At two in the afternoon we were on safe ground once more.

Near the entrance of Wild Cat Hollow we camped again by the half-dry Stinson Creek on a dead end road. All night long two large trees rubbing together in the wind screached and groaned. When daylight came the exuberant singing of the Louisiana waterthrush dominated the creaking trees, while I sat on a log and practiced sketching a flower. My lines never seemed to get the exact pitch capable by the shrieking trees, nor could my attempt to bring a flower out of the paper approach the symmetric harmony of the song of the waterthrush. The flower as I knew it in early life, persisted in presenting itself as a entity with all its impressions of beauty and accompanying memories of places and people. Now it must be transformed to a set of separate parts according to my knowledge acquired in later years.

In Cooper County we met Mr. S. and son in their field. He remembered X.Y. from the time he had collected there ten years before. Mr. S. said that his grandfather had killed 400 deer near there in Civil War times.

By using both hands and feet we crawled up the bluff south of the river. After all our strenuous efforts we found no plants of interest there. As one looks up from the bottom, bluffs at the top of the hill can be very deceiving. Having come to a place where it is no longer possible to go forward, as in one case a perpendicular bluff rising right out of deep water; or where going back over a very trying area to find a way to the top is scorned, the climb straight up to the top seems to be the line of least resistance. It looks easy from below, however the slopes of loose chert where you slide back one step out of two; or the thickets of briers and brush which looked from a distance to be smooth and easy, prove to be treacherous on close acquaintance. You crawl under, go around, or push through the tangles, avoid holes, and slippery places near the bluff, or climb over logs and large boulders not seen from below. Now you face the perpendicular bluffs which from below you thought to be four or five feet high. They loom straight up in front of you ten to fifteen feet high at the least, with nothing to cling to and no opening to get through. When alone, if I fell, who would know where to look? Not one chance in a hundred would bring X.Y. back that way. Not one day in a hundred would anyone else come that way. On this particular occasion I went up unconsciously collecting cryptogams on my clothes as I clung to the rocks. I breathlessly vowed never to do it again.

That night we thought we were to be entertained when the hoot owl began with notes resembling the opening bars of the opera "Carmen." That morning, when the water in our water jug was down to the last drop, a Louisiana waterthrush kept singing, "Brush your teeth clean, your teeth clean." A Bewick's wren announced, "I'll build my nest in a cedar tree. I'll build my nest in a c-c-cedar."

In Venezuela birds speak Spanish. One bird sings from dawn to dark, "Qué particular aquí, qué particular aquí?"—What's special here? I tried to tell him that there were bananas, oranges, seeds, and sugar water, but he obviously cared only for insects, since he never visited the feeders. After our breakfast at this camp I walked one-quarter mile with X.Y. and brought back water for the camp.

We were trying to find Chimney Rocks on the LaMine River, but two men who had lived to old age within a mile of them never had heard of such a place, though the map showed their location. The rocks we had visited years before looked familiar, but the road over which we had then driven existed no more, therefore we had to hike from Highway 40 to the bluffs and then climb them to get the picture we wanted. Mockingbirds were singing this far north.

My progress was often slow. When I collected cryptogams, I was always on my own. The closest cooperation we could make was to set a time to return to the car. Much time is necessary to scrape lichens, algae, and mosses from rocks, bluffs, trees, or soil. Sometimes 15 minutes are required to get a thimbleful of algae. After a full morning of such collecting I returned from the top of the bluffs down one ravine and reported to X.Y. that he had missed the best display of bluffs covered with Maidenhair and Narrow-leaved Spleenwort Ferns. He returned by way of another ravine and reported to me that I had missed the great display of ferns.

At Van Meter State Park we hiked over the 500 acres and found Showy Orchids (*Orchis spectabilis*). This orchid with two leaves resembling those of the Twayblade orchid has much more attractive flowers with upper lip of lavender-and-purple and lower lip of white. It is one of the most beautiful of the 31 native species of orchids in Missouri. Harrison and Nodaway Counties in northwest Missouri appear to be blessed with orchids. Near Pawnee we hiked through the woods and found 23 Yellow Lady-slippers, 50 Twayblade orchids, and several green orchids.

One might expect orchids to be more plentiful considering, first, the fact that the orchid family has worked out, with an apparent genius that man could envy, the most original schemes and diversified systems for compelling insects to cross fertilize them; and second, the fact that one seed capsule may contain over one million seeds. No other plant family has worked out such clever systems of structures for each individual genus to adapt to one particular insect best fitted to cross-fertilize that particular orchid. Some orchids actually resemble the insect they wish to attract.

When the number of orchid genera in the world are taken into account it would seem that the family would have run out of patterns before each genus was modeled. Self-pollination is not known in the orchids.

In the case of the Showy Orchid X.Y. had just collected, a female bumble bee is the one that ruptures the membrane containing the pollen masses and carries the pollen on her bald face to the next flower.

The Yellow Lady-slipper uses a different trick to get help from the bees. The bee enters the large upper opening, but cannot get back out that way because a valve-like structure that opens only one way, the way the bee goes in, compels him to find another way out. He is forced to pass over the pollen masses to reach the back door out, carrying the pollen with him.

The genus *Habenaria,* including the Yellow Fringed Orchid, has a nectar tube two inches long to attract the sphinx moth. After dark the beautiful Yellow and White Fringed Orchids are visible to the night-loving sphinx moths with long sucking tubes which they keep curled up in a coil when not in use.

Even though the orchid seeds become fertile with the help of the insects, these, the tiniest seeds in the world may fail to grow. One million seeds would not fill a thimble. There is no food stored in the minute seed as in the butterbean or acorn, therefore it must fall on a fungus where it can feed. There is a symbiotic relationship between the orchid seed and fungus. They need each other just as the orchids and insects need each other. In addition this very delicate germinating seed must get the right moisture and light to give it a start in life. The wonder is how any ever get started in the wild.

Anyone who has any conception of the struggle an orchid must undergo to perpetuate itself would leave any member of this family in its natural habitat.

That night we camped on the East Fork of the Grand River. The owner of the farm came along while we were eating supper. He was trying to recover from the tornado which had destroyed his home and the town of Worth the year before. He said he went into his tornado cave, but even then it felt like the tornado was pulling his ears out. He felt just like the trees looked after the leaves were all stripped off. Air pressure reduced to zero in the cone of a tornado permits everything to expand resulting in such freaks as undamaged straws in telephone poles. When the air pressure on the pole is removed and a crack appears, the straw enters. Then when the pressure returns as the cone moves on the straw is imprisoned as the pole returns to normal size.

West of our camp I found 50 Showy Orchids in a ravine. We were now hunting for the Small White Lady-slipper (*Cypripedium candidum*). We found many Yellow Lady-slippers, but no White ones. X.Y. thought he should find some near the base of the slopes. After we had walked over hills and bottoms almost all day, I started back a little earlier than X.Y. To avoid climbing up over another big hill, I kept to the same level about halfway up to the top. There to my great surprise was a colony of Small White Lady-slippers. The white of the slipper resembles white china more than anything else I can think of. I called repeatedly to X.Y., who was crossing the top. He finally came to find out what all the shouting was about. What a surprise to find the Small White Lady-slipper growing at a much higher elevation on the hill than expected. Often we have a preconceived idea of where a plant might be found, or of what habitat a plant would prefer, all based upon the only place it had been found. Sometimes to our surprise it excels in what seems to be a different type of location. The blue Bottle Gentian usually found growing in low damp places persisted and spread widely in the semi-shaded edges of an upland prairie where one would not expect it to grow and multiply well, if at all. In this case it was later revealed that the few inches of top soil threaded with mycellium was underlain by a hardpan clay mixture which prevented drainage long after the lower slopes of rich loam had dried out.

Though it sometimes seemed that X.Y. simply liked to challenge the car over the worst possible roads, he was merely trying to save the time of walking and carrying the heavy press. On this day in Worth County we started out on an abandoned, narrow road, washed out on each side. There were no rocks for road building this time as in many former times when ruts became too deep. Rain was threatening and thunder rolling. Finally we came to a washed out bridge in front of an abandoned house. Since we could not cross or turn around in the road, we pulled into the barn lot. There was an old water pump, and the day was drawing to a thundering close. X.Y. tried the pump hoping for water for washing, if we decided to camp there. Just then the new owners of the farm came along from Isadora to inspect the house they had just bought.

They suggested that we use the house that night if it rained. They thought it would be hard to find anyone willing to camp in the rain even if paid to do it. They left after advising us to pull the door shut tightly when we left the next morning, and to follow the meadow-divided Platte and Grand Rivers, if it rained. She was a biology teacher and interested in the work we were doing.

Near the house X.Y. saw a colony of plants that he did not recognize and dug up one plant. He thought he had something new until he smelled the root—cultivated Horseradish, a joke on him. The threatening rain did come down by dark, and we were happy to have a shelter.

On the road homeward through Scotland County we took the road to South Fabius River—called Fabby River by the natives who did not recognize the name Fabius. There we ran into pure dry sand one foot or more deep over a stretch of one-eighth mile. A farmer with a tractor had to come and pull us out. Getting stuck in dry sand is worse in some ways than getting stuck in mud. That night we witnessed a display of northern lights from eleven to eleven-thirty, one of the best we had ever seen. From my list of impressive mental trophies I pulled out the best and most mysterious aurora borealis for review.

Returning from the Olympic National Park in Washington State, we stopped in Montana for a picnic supper. It was a near-frost night in early September. We had finished supper at the edge of an open pine forest. I happened to look up at the bright moon in a cloudless sky. To the right of the moon at a distance of what appeared from earth to be six feet, I saw a shell with fluted edges about the size of the open hand. As we watched, colored strands began to emerge and wave like ribbons descending toward the earth for all the world, like colored ribbons floating from a Maypole. They spread in shimmering strands forming a cone. We seemed to be in the center of the cone. We twisted in every direction until our necks were stiff. After having reached the ground, the streamers gradually dimmed and disintegrated. At no time before or since have I seen an aurora borealis originate overhead and descend.

A last minute hurried hike over some bluffs resulted in finding *Polypodium virginianum,* an Ozark relic. Thus ended a fruitful May expedition.

On this 2,600-mile plant search through Missouri, from May 13–23, we did some visiting at the homes of botanists. While at Dr. G's in Liberty from Friday to Tuesday, we and others from William Jewell College made small expeditions out to swamps, Platte River bluffs, and points along the Missouri River. On the last day we went to explore a virgin woods 16 miles east of Liberty where we found the yellow-flowered Columbine. I found a red alga on damp rocks to add to my cryptogam collection, and X.Y. found a Cedar new to the county. The clear yellow Columbine is found only in two counties, Clay and Buchanan, while the red one is found in almost every county in the state.

From Dr. G's home we drove to Lafayette County near Odessa. While I made camp for the evening, a six-foot blacksnake kept bothering me. Again and again I chased him back across the road from where we were to eat and sleep. While we ate supper he sneaked back to our side. Or was it his side. Perhaps we were the trespassers. He seemed to have more persistence than fear—moving slowly away when driven, but promptly returning.

Six miles south of Pittsville the following day we found white Closed Gentians (*Gentiana flavida*), two kinds of *Delphinium, Arenaria, Heuchera,* and *Ophioglossum*.

At Knob Noster State Park we encountered camp problems again where my warm, green jacket walked out on the back of one of the high school students picnicking there. Our crackers also proved to be better than theirs. There the students were putting on such a wild display of letting off steam, that a man who entered the park later—thinking they were ours—asked if we were giving a circus.

On the South Fork of Black-Water River in Johnson County we found lovely mats of light purple *Astragalus distortus,* a lovely plant for rock gardens. While watching a summer tanager and other things of interest, X.Y. got too far ahead of me, and I tried to bring him back or halt him by yelling, "I'm lost." He did finally wait, and we crossed over a dozen canyons and deep ravines, then returned over the top across many shale barrens entirely bare of any plant life even in May.

Next we drove toward Dresden and stopped along a railroad to inspect the prairie flowers: showy Primrose, deep rose-colored *Phlox,* Shooting Stars, acres of blue *Tradescantia ohiensis,* Yellow Puccoon, *Senecio, Delphinium,* and hundreds of Downy Gentians (*Gentiana puberula*) which would be making quite a show of deep blue-purple with red-purple outer veins in September and October.

We rambled around over Pettis County and found no rare plants, but tried to read into the landscape the reason for the prosperity of the county 75 or more years ago, evidenced by so many 8- to 12-roomed, two-story houses; and the reason for the decline of prosperity which left the big old houses unpainted and deserted. With self-interest predominating; with no idea of any obligation to the land or the future; with plenty of virgin land in sight, a farmer could discard a declining farm as easily as a woman could discard an old dress for a new one.

The one-time gift of the land showed in the large homes and barns. Now the vacant, unpainted homes, ghosts of the past, matched the eroded, over-grazed, baked clay soil worn to the stage where it could scarcely support stunted weeds. Man makes quick and violent changes in a landscape within his own lifetime. In this county not man, but men had been guilty of ecological blindness.

Where we stopped near Sedalia X.Y. got ahead of me as we crossed a field and a clear creek, which I was trying to cross without getting my feet wet. I heard him "Yoohoo" on top of a bluff. I pulled up the steep, rocky bluff in the direction of his voice. When I arrived at the top, from where his voice had come, it now came from a point some distance south of me. I went south in search of the voice and when I reached that place he was gone. By then I heard him call down at the road from where we had started. I had made a home run for nothing. Finally we both reached a glade where he found the plant he was searching for.

At Pershing State Park, where we arrived at dark to cook a good supper of pork chops, potatoes, and corn in their nice shelter house, we found the building had burned down. Our good supper amounted to a cold cheese sandwich, tomato juice, and an apple. When X.Y. crawled into his sleeping bag there was not a cloud in the sky. When I got my bed prepared there was not a star in the sky, but I was prepared for rain. At two in the morning X.Y. had to move into the car, while I slept out under the plastic.

At six-thirty the next morning we had our proposed dinner for breakfast. Pork chops, potatoes, corn, coffee, and chocolate fortified us for the day. There was time to prepare such a breakfast as X.Y. had to put all his specimens of the previous day into press.

With the rising sun, clear and hot, came the mosquitoes. They came up in drifts from a low swampy area in the woods. To prevent a special seasoning of mosquitoes it was necessary to keep lids on the cooking food. In all our experiences of camping no other place excelled in mosquitoes, day or night. Except for the mosquitoes, it appeared that we had reverted to leisure as we lingered around the camp until ten o'clock.

Near Browning in Sullivan County along the sunny banks of the railroad we made a collection of Spiderworts (*Tradescantia bracteata*) four inches high with colors ranging from white, pink to rose, rose-purple, lavender, violet to blue-purple; a riot of color for a rock garden.

Near the end of the day we came to a dead end road where one of those large, old abandoned houses stood. We made camp there along an old neglected *Rosa rugosa* hedge. When neglected this rose spreads but continues to make a display of large, single pink roses. The lawn of tall weeds showed no sign of trampling. All the windows of the house were boarded up except one near the back where I later noticed a length of curtain hanging inside. After X.Y. had gone away on a last minute plant hunt, I saw a triangular space open between the pieces of curtain. The space had not been there earlier. I thought perhaps a long-time owner, finishing life with the old house, had reduced the living quarters to one or two rooms in the big old home.

Near dark a very old lady, stooped and bent, came from a rear door and went to an old barn to milk. If the poor woman were frightened at her visitors, she must have been more so when X.Y. came back just at dark yoohooing like an Indian demanding supper. He still thought the house was vacant. We hoped the poor old lady slept that night. No doubt she was ready for an emergency, unless she had the ability to analyze the situation and sense our innocence. At any rate we each kept our distance.

Some rain came again while we ate breakfast. The only rainless 24 hours of the 11 days we were out was the preceding Friday. Our equipment now included mold and mildew, plant specimens we neither needed nor recorded.

On the road home at Chenoa, Illinois, where we stopped to get gas we found a rear tire low. The man who had fixed the tire as we started on the trip had left the paper inside the patch and the tire had held up for 2,600 miles over all kinds of roads.

CHAPTER VI

MAY

Floods

There had been floods on the Meramec River a short time before our visit to Missouri on May 16 for more extensive work on the "Flora." The banks still covered with mud left from the flood, furnished no place to sit and eat lunch.

We arrived at Botany Hill, our first destination, at the home of our botanist friends, the B's, just before dark, and spent the night in their charming 117 year-old home on top of the hill. Before leaving the following morning we explored his nature preserve rich in native and transplanted flora.

Next, we drove over highways 61 and 50 to see the pink Azaleas (*Rhododendron roseum*). Near Castor River bridge, the Azalea display was at its best. Across the open, rocky field on north- and south-facing slopes, and in shallow ravines hundreds of well-pruned Azalea bushes were covered with pink to rose blossoms. Here in full sun they flowered better than in the woods. The taller unpruned, widely-branched shrubs in the partly shaded woods put on only a few flowers as compared with those in the pasture. We walked in to photograph some of the best specimens. The fragrance somewhat resembles that of the Japanese Honeysuckle blossoms.

This field of Azaleas was obtained, developed, and established by the Fredericktown Rotary Club in May 1960 as Spiva Azalea Park. It was dedicated in memory of George Newton Spiva who was interested in preserving the natural beauty of the native Azalea. Each year for 50 years during Mr. Spiva's lifetime, he let nothing stand in his way from coming to this spot from his home in Joplin during blooming time in the spring to enjoy the beauty and fragrance of these flowers in their natural state.

After photographing the Azaleas for color prints, we reluctantly left for Cape Girardeau County. There X.Y. explored one of the richest virgin bottom woods in Missouri where he found White Trillium (*Trillium flexipes*). The large white flowers of this Trillium hinting of sweet pea fragrance hang down and is partly hidden by the wide leaves. There were also Bluebells and Cowslips. And since the word "Cowslip" in gardens, painted on china, and used as prints on cloth or paper may mean: *Caltha palustris,* the Marsh Marigold; or *Primula,* a Primrose; or *Mertensia virginica,* the Bluebell, it is advisable to follow "Cowslip" with *Mertensia virginica.*

Snakeroot is another word that snakes itself into six different families. Hours may be spent trying to locate the desired information when only the common name is known. Whereas, if the scientific name of the specific Snakeroot were known, the facts could be found in five minutes. Also there were Celandine Poppies (*Stylophorum diphyllum*). This yellow Poppy runs a good race with Bluebells in the capacity to spread, and makes a nice trio in the wild or in the wildflower garden with Bluebells and Trillium all blooming together.

In these rich woods X.Y. found Adam-and-Eve orchids or Putty Root (*Aplectrum hyemale*). The habits and bulb-like roots of this orchid furnish more interest than the rather dingy, small flowers of yellowish-brown and purple color. In the fall one broad, dark blue-green leaf comes up directly from the center of a corm similar to that of a crocus. The two corms underground, Adam-and-Eve, are attached by a thin root. In the spring about the time the leaf dies down, a flower stalk arises from the side of the corm. In the meantime a third corm is forming, attached by a thin root to number two in a straight line. During the summer when the flower and leaf both disappear, the corm producing them dies also, leaving corm number two to produce the new leaf in the fall. Corm number three tags along, and next year produces number four, still in line.

Corms of this Putty Root orchid have been ground and mixed with water to make a paste or glue suggestive of the tenacious habit of the successive corms.

From Jackson to Cape Girardeau on Highway 61 we drove for 18 miles through a lovely rose garden which someone, or a group of people had planned, executed, and maintained.

In New Madrid County we walked for sometime over shrub-covered sand dunes and found a lavender Clematis, Yellow and Orange Puccoons, and Passion-flowers. X.Y. found one plant he did not know, causing great excitement.

Everywhere on the country roads P. and Y., water stood over everything on each side of the road throughout this swamp country. Houses showed water marks as high as the tops of doors and windows. Most buildings had been abandoned. Near Latorge X.Y. tried three different roads to reach a wooded area surrounding a lake. Each road led to a dead end. At New Madrid they told us that high water had cut off the approaches. Spring evidently was not the time to botanize in this swampy area. Strangely enough, soon after leaving New Madrid we drove through high winds and a dust storm.

We were headed for Poplar Bluff, to Cane Creek, a place six miles south of there to spend the night. While we were eating supper a man and woman who had been fishing came and told us not to stay there if it started to rain. They were living on top of the hill in a small temporary house since their home had been blown to a thousand bits by a tornado. Right then it looked as if a terrible electric storm was near.

They pointed to cornstalks and other rubbish in the trees high over our heads and said that the creek was subject to flash floods that rapidly raised the water that high. They could not accomodate us in their small house, but advised us to go to higher grounds. The flood trash over our heads seconded the advice, and we voted to leave for higher grounds.

Almost before we were through eating it began to rain hard. The road on which we intended to continue crossed the creek, therefore it appeared advisable to get across the creek and out of the lowlands to a hill beyond, before water cut us off. Without delay we packed the car again, drove on to a safe elevation, and slept in the car while it rained all night. The rain continued through most of the day as we drove toward Rolla. We were getting a wet reception everywhere.

The only place we could find to camp that night was a Camp-ground School. It was near a nice creek, but the stream had been up the night before and the banks were too muddy. I placed my cot in the middle of the graveled parking area of the school, and X.Y. placed his sleeping bag near the foundation of the stone church and school.

During the night I dreamed that something was stinging my left arm. A very sharp, hot sting woke me up. Then I felt something heavy on my bed near my shoulders. It was trying to nudge under the covers. Since it was a cold night I thought perhaps a cat was on my bed, however there was no house in sight. I gave the covers a flip and said, "Scat." Something dropped to the ground and I could faintly hear it move over the gravel. I went back to sleep and forgot about it.

The next morning X.Y. came to where I was preparing breakfast and showed me his arm. Something had bitten or stung him on the forearm. It was causing considerable pain. There was one red mark in the center of a black circle about two inches across and the arm was somewhat swollen. We could make no guess as to what would produce such a queer mark. I related my dream, which he thought was a strange coincidence.

After breakfast when the sun had warmed the air somewhat, I took off my sweater and there on my left forearm where I had dreamed I was stung was the same kind of mark X.Y. had on his arm, one red spot surrounded by a two inch black circle, though I had no pain and very little swelling.

We were bitten by the same kind of enemy, and both bitten between the wrist and elbow as if some disturbance during sleep had caused us to throw the arms out to ward off something. Now was the weight, about that of a cat, on my bed which I could hear move away coincidental with the sting or did it cause the sting? Our guess was a scorpion or black widow spider, but neither could have had weight or have made enough sound to be heard when running over gravel. In my location in the center of the graveled area I was not so much exposed to scorpions or black widows as was X.Y. on the ground by the stone building. The weight and sound could have been that of a snake, but there was only one red mark. We do know that scorpions get around at night. The spots persisted throughout the day. The mystery was never solved.

Two frogs at this place kept up a duet. The one in the roadside pool near us called loudly to another in a pond far across the field of alfalfa. Both were reluctant to move, at least during the hour I was there. They were still broadcasting when we left, though tortoises were moving. Tortoises are always wanting to cross the road. Two half-grown ones taking a sunbath on the warm gravel had decided to go across the road and ran into difficulties in a deep rut. One turned a somersault into the rut and could not get back on his feet. I lifted both up to a place of safety where they went on into the woods.

Floods and rains were defeating us. We were doing little more then driving to find some place not flooded. Both the Osage and the Missouri Rivers were swollen. We finally took to the hills looking for the large White Trillium, and Hepaticas on north-facing Missouri River bluffs. We did not find the Trillium.

At one place we came to a bluff which X.Y. partly dropped down. I like to see where my feet are going to land when I slip over a bluff.

I had no parachute like the leaf-cutting ant that found his advance cut off four feet above the ground by an approaching spider. They proceeded toward each other until about six inches remained. Then the ant suddenly turned loose of the rose cane and using his thumb-sized piece of rose leaf as a parachute dropped safely to the ground. Since this emergency obviously demanded prompt and individual solution it upset my previous idea that the ant is a robot.

If I swung down a bluff, there would be no one there to assure me that the bottom was there and safe for landing, as I was once assured when I turned loose to slide down the side of a glacier into the stark blackness of its interior. At any rate they told me the bottom had been there just a few years ago and the glacier had not melted inside for ages.

At this time, instead of sliding I decided to walk along the top of the bluffs where I came to an old, abandoned cemetery in the woods. *Iris, Yucca, Narcissus,* Myrtle, *Spiraea,* and other cultivated plants sprawling over sunken graves without markers presented a strange setting in the wild woods among weeds, grass, and shrubs.

After lunch we drove to another bluff over the Missouri River. X.Y. went west and I went east where I found carpets of Blue-eyed Mary and *Isopyrum,* both new records for the county.

That evening we went to the Lake of the Ozarks to spend the night. There on the shore of the lake above the high-water mark, city people go to spend a weekend or vacation surrounded by as much of the city as can be condensed in a narrow area around the edge of the lake. Neon lights contest with moonlight across the lake. Jukeboxes, radios, and televisions blare out noises amplified in contrast with their noiseless setting.

Bugs, little bugs and big bugs, from no-seeums to June beetles, battled our nerves all night. We left the bugs and noise and other city attractions, and as fast as possible the next morning headed for Wheatland.

At Wheatland in Hickory County we looked over a glade sporting swaths of Indian Paint Brush. Plants of a natural prairie attracted X.Y. like a clover field attracts bumblebees. We were about to pass one. Since time was running short and the prairie was safely enclosed inside an eight foot woven wire fence, I decided to sit on the running board of the car on the shaded side and write up some daily notes.

X.Y. was gone some time. Returning with his hands behind him looking as pleased as a cat from an open milk house, he said he would give me $1,000 if I could guess what he held behind him. I was becoming independent from the $1,000 bets won by a sixth sense on this trip. He was certain I would guess a rare flower.

I said, "Katydid."

His eyes popped then he asked, "What kind of katydid?"

"Pink," I promptly answered.

He almost dropped the katydid. It was pink, full-grown and alive. Up to that time I never had seen or heard of a pink katydid. Later, in that vicinity I saw others on bushes.

Going farther south we stopped to inspect some rich bottomland along the Pomme de Terre River at a place where there had been no fire and no grazing. In that lush area we found new records: Yellow Lady-slipper, and Coral Root orchid (*Corallorhiza Wisteriana*), a strange degenerate or beggar in the orchid family according to some. It has no leaves for making its own chlorophyll and therefore lives on other roots as a saprophyte or parasite taking its nourishment from the roots of the host plant. The dull purple stem, flowers of yellowish-green with the white lip dotted by lavender or purple are not very showy. Rudimentary leaves, no more than scales on the flower stalk, might lead one to suspect that the plant once had leaves, and through generations had developed a way of getting food with less effort. Should a plant or animal that shows evidence of having changed from a state of self-service to one of being served be called degenerate?

Thousands of tropical orchids producing large showy flowers have elevated themselves to habitats above the ground. Though they have leaves they have managed to use air, moisture, and a minimum of dust, and have developed the most complicated and varied systems for fertilization of any family of plants. These plants can scarcely be called degenerate. Perhaps upward evolution is a better term than downward degeneration for the Coral Root.

Yellow *Baptisia sphaerocarpa* was found and photographed three miles south of Cairo in Randolph County west of Highway 63. In Adair County south of Yarrow we walked for a mile on the sandy bottom of Sugar Creek, wading in the clear warm water. Leaving the creek we walked for one and three-fourth hours along slopes rich in leaf mold and plant life, and found 41 large Yellow Lady-slippers.

Monday night we had a camp fire steak supper on Brush Creek in Schuyler County. The first man who stopped while we were eating supper thought we were hunters; the second one thought we had car trouble. X.Y. had a headache and nausea all night. I also had a groggy head the next day, possibly the results of the former mystery bites.

Going eastward, browsing along the way, we finally left Missouri on May 23 for home having piled up 2,056 miles on Missouri rambles in the eight days out.

One year and four days later we were in Missouri again for the same purpose. We made camp on Castor River southeast of Fredericktown. People were hoeing young cotton plants near where we were looking for a lost prairie.

Since X.Y. had found the Pond Berry Spice Bush, new to the state in the fall he wanted to visit the same swamp on Mr. H's place in the spring when it would be in flower. We made three trips to the house before finding anyone home to give us permission to go into the swamp.

Later at another place on a swampy road we met an elderly man looking for catfish in the ditches along the road. He said that the day before when he was at the hospital with his son (who had been injured in a car wreck), a man was brought into the hospital with a rattlesnake bite on his heel. "Children here wear shoes only to church," he said. "I can go squirrel hunting barefooted, my feet are so tough."

X.Y. suggested, "That is a good way to save shoes, if you watch out for snakes."

The old man then asked, "What can you be doing down here? I didn't think you could get anyone from Chicago down here into these swamps."

We left the man probing for catfish and headed again for the Bogbean place in Mr. G's swampy meadow. After some terrible roads and many crossings of graveled-bottom creeks, by sheer luck, we arrived and found that the Bogbean still survived.

That night, when we camped along a small creek some form of animal life kept us guessing. I thought the cries throughout the night came from a water bird. X.Y. said it was the whining of a small dog, but later he said it was a kitten. We never saw it and never knew.

Decoration Day we left for Green County and drove steadily until six-thirty in the evening since there would be no camping out that night. We were surprised to find the country around Springfield so dry and bare after all the floods we had been through. There were very slim prospects for botanizing.

Near Hoberg in Lawrence County, we spent the night at the home of friends, the F's, who raised thousands of turkeys, and had about 150 peafowl that screamed, "Help, help" during the night every time anything moved. The next day they strutted and posed for their photographs.

The last day of May we spent on the Pomme de Terre River exploring river bottoms and bluffs. To get acquainted with nature, not in the sense of acquiring facts, but in the sense of perceiving, one must go alone or in the company of only one other equipped with the same perceptive development.

The sentence, "I heard the song of a winter wren," when heard in snatches above the crunching of 40 feet over a rocky trail in the Smokies could mean little more than just that. While the same sentence the following day, when heard alone on the heavily forested trail to Le Conte Mountain, meant not only the song complete with every note of one wren, but of apparently several wrens on different sides of the mountain in succession. One bird took up the song at the last note of the preceding one without a pause to hear the first one's echo. It continued on without break as if the same bird had moved across the mountain with lightning speed.

Pausing on the trail to listen to the continuous song coming through the foggy shafts of sunlight disbursed by tall pines, one could climb and descend with the exhilarating notes of the song. Not a sound competed except an occasional drip of condensed fog or the softened plomp of a Rhododendron flower on the damp trail.

Here one could say, "I heard the song of the winter wren," and mean the full melody of song accompanied with the entire setting of the silent, damp, and cool mountain trail with the concomitant fragrance of damp moss, ferns, and pines.

The wren song as well as other non-violent sounds of nature such as wind through the pines, rustle of leaves, running water, and lap of waves all tend to be restful and soothing to nerves. Such sounds seem to be musical. If their vibrations were recorded on unbound substances such as sand or powder, or recorded on graphs, they would show only musical patterns, a repetitious design of harmonious figures, curves, crests and hollows in patterns pleasing to the eye as well as the ear. These musical sounds of regular patterns are in keeping with the rhythmic functions of the body—the heart beat, breathing, peristalsis, and so on, hence soothing. When the metronome of the body is jarred from its natural rhythm by noise, unhealthy reactions take place in the body—nervousness, irritation, or even sickness.

Noise is the irregular jumble of pitches with no pattern. It is said that children and savages love noise. As for us at the end of a hot day we found a good place to camp on the banks of a rock-bottom creek where we relaxed from the complexities of modern existence. In addition to plant collecting, we had collected many mementoes not to be hung on walls nor placed in drawers or cases. We slept peacefully while a bullfrog croaked all night ending another May.

CHAPTER VII

JUNE

High Tide for Flowers

The month of June is a high tide for flowers as, at this time, spring flowers overlap summer flowers, more or less, according to seasonal conditions, or the distance explored to the north or south in the state.

Though highway 66 covering most of the distance from our home in Illinois to Missouri is stranded high and dry, and for the most part clipped and pruned to monotony, we scarcely traversed a mile without seeing some plant or bird of interest. One year 20 round trips were taken covering every blooming month of that year, yet each trip furnished a different interest. Each unusual thing noted, registered more to us than the vista from an observation point at New Found Gap in the Smokies. A man who drove up quickly with his family was overheard to say, "Just another bunch of hills. Let's get going."

On this June 7, on our way from home to follow another plant quest for the book now growing larger and larger, we hurried along to get south of St. Louis by lunchtime. About 15 miles north of Sullivan we stopped to examine the flora on a sunny strip between the highway and the Pacific Railroad and found Depford Pinks (*Dianthus Armeria*), which had escaped from gardens to battle the weeds along the roadside.

Here also bright patches of Lead Plant (*Amorpha canescens*), and Hoary Pea (*Tephrosia virginiana*) demanded attention. We stopped to admire these and make a brief search for others associated with them. There exists other flowers more showy, more plentiful, and more easily grown than *Amorpha* and *Tephrosia*, yet there are few others that can form a similar synapse between plant and person. Their rarity; reticence; and their ability to extract nourishment from barren soil, and chert of open woodlands and prairies, demands admiration. Their ability to transform those inert elements to beauty and life can still the heat of a summer day.

Coming upon them on solitary hikes a mutual common denominator is sensed. The mood of a person at the first acquaintance with a particular flower often repeats with each following encounter. No one doubts that the fragrance of a flower revives memories connected with its fragrance from another specific time. In the same manner the sight of a flower revives feelings probably only subconsciously felt at the time of the first acquaintance with that flower.

These two leguminous plants, often growing together, each with downy grey leaves reminiscent of desert foliage adding exactly the proper base for the flowers, were found here along the railroad right-of-way. The Lead Plant from one to two feet high with pinnately compound leaves of from 31 to 57 grayish, half-inch long leaflets, stand erect or arch, forming a lovely background for the six to twelve inch long flower stalk of bright blue-purple, tiny pea-like flowers in a compact mass along the spike. Bright yellow stamens sticking out beyond the petals complement the purple petals. It needs full sun to be its best. Though it can be grown from seeds, it stubbornly resists transplanting. Its special requirements are not known.

The other legume of equal favor, and as difficult to propagate is the Hoary Pea or Catgut (*Tephrosia virginiana*). *Tephrosia* meaning ash-colored refers to the leaves. Seven to 28 woolly ash-colored leaflets make up the pinnately compound leaf. Flowers are borne on a raceme at the tip of a branching flower stalk one to two feet tall. Individual flowers are somewhat smaller than the commercial sweet pea. The banner of the flower is cream-yellow, and the wings are rose to purple, while the keel is a greenish-yellow tinged with rose. This lovely combination of soft, pastel colors makes a nice contrast with the bold colors of the Lead Plant. At night these plants would scarcely be recognized with the leaflets folded in sleep. Both legumes seem to thrive under adverse conditions, yet must require some unknown element existing in those sparse settings.

Growing in similar localities we found New Jersey Tea (*Ceanothus americanus*) which grows from one to four feet high in clumps. Spherical or blunt cone-shaped heads arising from the tips of the stem and axils of the upper leaves are covered with tiny creamy white flowers which appear foamy or plume-like from a distance. The name "Redroot" comes from the red root of the plant. In Revolutionary times soldiers made tea from the dried ovate, pointed leaves. Experiment supports their good judgment that the dried leaves really do resemble the tea of commerce in flavor.

Elderberry (*Sambucus canadensis*) were beginning to bloom. This tall shrub of the Honeysuckle family is known better by the loose clusters of shiny black-purple fruits than by the lacy, creamy flower cluster. We have experimented with most known uses: tea, flavor for pancakes, pie and puddings made with ripe berries, corn starch, and lemon juice. One of the most common uses, however, is Elderberry wine. A story as told by a botanist who was exploring in Mexico goes: He found it necessary to stop for the night at a Mexican farmhouse. While walking around with the farmer in the evening the botanist saw that the chickens were going to roost in an Elderberry bush. He remarked to the farmer in Spanish, "I guess you would call that tree an 'arbol del pollo'," meaning "hen tree."

The farmer answered in all seriousness, "No, señor, we call it *Sambu-*

cus Mexicana!'' The botanist from then on had a higher respect for the Mexican knowledge of plant life.

Along the highway south of Cuba we found Rattlesnake Master or Button Snakeroot (*Eryngium yuccifolium*). From the silver-green yucca-like leaves at the base of the plant, a stem arises about three feet which branches. At the tip of each branch a stiff ball about an inch in diameter is covered with whitish flowers. The plant has been used as a remedy for snakebite.

Sensitive Brier or Bashful Brier (*Schrankia uncinata*) were boldly running along the banks of the right-of-way in full sunshine. The balls of pink-purple flowers occurring along the rambling stems are pretty and soft to look at, though not to be picked and are very treacherous to walk through since the stems are thickly studded with curved spines. The pinnately compound leaves of numerous small leaflets are very sensitive to touch and close up at the touch of a tip to the leaflet, hence the name ''Bashful Brier.'' This plant refuses to be transplanted or to be grown from seeds planted by man. Its secret requirements have not been learned.

We drove on to camp at Jack's Fork on Highway 17. Here we were reminded of a camp there years ago in November when we drove from the University of Missouri, to Chicago by way of Jack's Fork, a detour of 200 miles to take one picture of a stand of Juniper trees. A camp fire from snow-covered wood sizzled and sputtered, giving little heat at that time. We could have more easily warmed our hands in the stream.

But now in June beyond Mountain View the next day we found Carolina Larkspur (*Delphinium carolinianum*), Goat's Beard (*Tragapogon pratensis*), and Yellow Dog Fennel or Bitterweed (*Helenium amarum*). Besides having an unpleasant odor the Bitterweed contains poison. The name ''Bitterweed'' is no misnomer.

Butterfly-weed (*Asclepias tuberosa*) was voted high on the list of popular wildflowers. The root is also used for medicine as an emetic. Though very difficult to transplant, it grows readily from seeds. The brillant orange flowers persist for several days. The Common Milkweed found in most of the states (*Asclepias syriaca*) with its large close clusters of lilac or crimson-pink flowers is almost as handsome as the orange Butterfly-weed. The pods of this one are the most attractive. At times two adjoining pods resemble two light-green love birds perched side by side. When ripe the seedpod opens to spill out a cascade of silky wings attached to brown seeds. At one time we found one flower head which had trapped a great variety of insects. Besides the orchid no other flower has so well devised means for insect help in fertilization. The slippery surface where the insect lights skids him into a cleft at the base of the flower where his foot is trapped. In his struggle to get loose he picks up pollen which is carried to another flower. But he does not always get loose.

Near West Plaines we found Persimmon (*Diospyros virginiana*) of the Ebony family. It is one of many widely distributed species, though the only species indigenous to the United States. It is native from Texas to Florida and from Florida northward to Connecticut. In the Ozarks one often finds clones or thickets of trees produced by suckering. The cream-colored or pale yellow flowers are not very conspicuous. The staminate (male) flowers grow on one tree and the pistillate (female) flowers on a separate tree.

In early fall, cold nights and frost set little chemical factories to work inside the oval green Persimmon fruit to change the impossible, bitter astringent flesh to a sweet and fragrant form within the orange-purple to purple skin. Nature finishes these mature fruits with a coating of powdery bloom.

The fruits are relished by man and most wild animals. Jellies, pies, preserves, and ice cream are made from the fruits. The best results come from drying them in sugar. It is the favorite food of possums. Horses have been seen to bump the tree causing the fruit to fall then picking it up. In one case a large bulldog had the habit of making a long dash, hitting the tree head on to cause the fruit he desired to fall.

The black, red, or green seasoned heartwood of this member of the Ebony family is one of the hardest woods known. It is excelled only by Black Locust, Honey Locust, Dogwood, and Ironwood.

South of Birch Tree in Shannon County we inspected Heal-all (*Prunella vulgaris*). The blue flowers of this mint continue to climb toward the top of a blunt spike-like inflorescence from May to October. This habit of flowering is not the only generous one possessed by the plant. Having come from the Old World, it spread widely in space, in name, and in uses where Indians and herbalists used the Self-heal, Heal-all, Carpenter Weed, Blue Curls, Heart-of-the-Earth, *Prunella* as a cure for all ailments, especially for diseases of the throat.

Retracting our course northward through Winona on Highway 19, we walked some distance to an isolated over-hanging bluff and found three rare plant species. Northern Bedstraw (*Galium boreale* var. *hyssopifolium*) was one, found in only a few places in Missouri in Shannon and Texas Counties. It is a relic which has persisted on cool north-facing bluffs since the retreat of the glacier. At the same station we found Harebell (*Campanula rotundifolia*) known from only a few stations in Shannon County. It is also a relic from the ice age. This little blue bell can be grown in northern states where it is kept moist and shaded. At this place clinging to the cool moist sides of the shaded bluff, it is fairly safe due to the isolated and inaccessible spot. The third rare plant in this region is False Bugbane (*Trautvetteria caroliniensis*), the rarest. It is found in only a few places in Shannon County. It grows well from seeds in our northern wildflower garden.

Later along Highway 19 we stopped at a glade to look for possible new records. There we found the Rock Pinks (*Talinum parviflorum*) beginning to end their flowering season. These one-half inch wide, flat rose-colored flowers resembling small editions of the cultivated Rose Moss (*Portulaca grandiflora*) arise from mosses or lichens covering the thin soil of level glades adding bright bits of color to the grays and dull greens of the sparse vegetation carpeting the glades.

Traveling north to Franklin County we were caught in heavy rains. Ten miles southwest of Union we finally became stuck in the mud in the middle of the road for some time. The following day as it continued to rain we started out plant hunting regardless. It was not bad wading the rising Bourbeuse River as we hiked over to the bluffs beyond. But when we returned after searching over the hills and bluffs for a few hours in the rain, we were baffled by a roaring, muddy stream carrying driftwood. It was either wade across or walk a great many miles around to bridges to get to the car. We decided to wade. Holding hands (mine was shaking), we waded waist-deep in the swift muddy current. The water had risen over two feet in that time. I recalled a time on a former trip when X.Y. was trapped on the far side of a wide river that had risen while he was gone. He started to swim across and was having trouble with the strong current. I stood there swallowing my heart until his feet were firmly on land again.

We found Dutchman's Pipe (*Aristolochia tomentosa*) with flowers clearly resembling Dutchman's pipes, trailing over bushes along the stream. When the rain ceased we changed driers. A man passing called out, "You must be having a lot of trouble." Truly it did look like a mess when we changed driers.

Gradually we worked our way south to the Indian Creek area where we expected to spend some time. At Indian Creek we needed directions to a place not clearly marked on the map. We stopped at an old house and called out. An elderly woman who looked as weatherbeaten as the old two-story house, sat at the door reading a newspaper. Her skin was like tanned wrinkled leather, contrasted with the keen grey-blue eyes. Her salt-and-pepper gray hair, braided on each side was pulled back around her head. She had been busy whitewashing the picket fence around the lawn according to the lime-spattered shoes. Two young women peeped around the corner of the door, but did not venture out.

At our inquiry about the road the older woman came out to give directions. In answer to our query she said, "I ort to know where Piney Fork was. Henry'd know."

Just then Henry came down the hill carrying a scythe on his shoulder. I noticed a blue racer snake lying by the gate with its tail still quivering. Henry said, "Ma killed a racer. In May I killed a cottonmouth 42 inches long. I struck it on the head with a lucky strike when it reared its front parts.'' Since they had mistaken a blue racer for an enemy it is likely they mistook a harmless water snake for a cottonmouth.

By now Pa had joined the family saying, ''Good evening,'' at eleven in the morning. He said, ''The creek was plum over yander,'' pointing in the general direction of Piney Fork, ''but I plum forgot just how you get there.''

X.Y. asked how farming had been. Pa answered, ''We have our troubles. Indian Creek got up to the church yander, took our straw stack—never left a straw. You see that tree there. It got up there and took the new gate we put up just two days before. I thinked it out and guess we'd best to move.''

We started on foot to Piney Fork. Pa followed along carrying a rope. He thought we were fishing.

X.Y. showed him Snakeroot and told him it was good for medicine. But Pa said, ''That ain't what we call Snakeroot.'' He said, ''Put the berries of Pokeweed in alcohol for rheumatism. Fry the roots of Pokeweed in grease to cure dogs of running fits.''

Pa pointed out Indian Hemp (*Apocynum cannabinum*) and asked what it was good for. X.Y. showed him the sticky juice and tough fiber which was used by Indians for making fishnets, mats, and baskets.

Then Pa said, ''I guess I'd better plant a whole field of it. I've been cuttin' it out—every one I see.''

He took out his knife and cut down a tall Pokeweed. I remarked that the birds like the berries of the Pokeweed.

And Pa said, ''Let the birds eat the insects around here. There's a plenty.''

I said, ''I make Elderberry pancakes.'' Then Pa asked for the recipe.

He said, ''Ma uses dried Elderberry flowers to make a tea for babies. I guess God made every plant to be used for sumthin.''

When we came to the creek X.Y. and I waded right through, as we always wore tennis shoes which would dry quickly. Pa sat down and removed his shoes then waded cautiously across, hunting for flat rocks to step on. On the other side he sat down on the gravel and put his shoes on again. He had just finished lacing them up when he remembered, ''Hang it all. I left my rope on t'other side.'' He had to take his shoes off again and wade back for the rope. We waved goodbye and went on our way. We had met friendly people we would not easily forget.

Having finished our work in the Indian Creek area, we went through Potosi. On a glade north of Potosi some Missouri Primroses (*Oenothera missouriensis*) were still in bloom. Being one of the showiest wildflowers in Missouri it is well named. The long sprawling red stems and attractively veined leaves are of themselves attractive. The bright yellow flowers four or more inches across which open in the evening seem to glow with a translucent light by moonlight. The length of time the flower stays open the following day depends on the temperature for the most part. On cloudy days, or very cool days even though sunny, they remain open most of the day. Many other night-blooming flowers have the same habit. At the base of each long flower stalk, the three inch long seed pods with four wings, each one inch wide, insure a new crop of Primroses if given the proper environment.

We left Potosi going westward through Quaker near Delbridge, the ghost town, where we had a very nice camp on Cortois Creek. As we ate supper a red-headed fellow came by and after a short visit concerning the nature of the outdoor life we were leading, invited us to attend a little country church about one-half mile away. After our work was done, we started for the church being guided by the choir which could be heard for at least a mile. We arrived late, causing considerable neck-twisting as we entered. Being total strangers (where strangers seldom entered) one could expect nothing less.

The minister's sermon consisted of sin, hell, and heaven. It was delivered with a vocabulary of a few dozen words repeated over and over with little variation, which was especially directed toward the red-headed, teenage, big-eared boy who, with others, sat on the backs of the benches with their feet on the seats.

There was quite a din with the mourners all praying aloud for their sins to leave, and their spirits to come in. All this was accompanied by a rhythmic song played by two guitars and accented by clapping of hands and the stamping of feet by the boys on the benches in back of us. We behaved with decorum.

Cold winds which struck us in the back from the open door started us back to camp promptly with the last "amen." We left it to our red-headed friend to explain to others that we were his guests.

The following day we went through Berryman stopping only once about seven miles from Berryman to get a specimen of the delicate blue-flowered *Heteranthera limosa* which was growing in the mud at the edge of a pond.

Sunday night we spent in Steelville in Crawford County where we left early the next morning to explore around Fishing Spring and Meramec River. Around Wesco we found Ginseng (*Panax quinquefolium*). The shiny red fruit cluster arising from the junction of the three leaves is more conspicuous than the creamy white flower cluster. The large fleshy roots used especially by the Chinese for medicinal purposes have been dug to supply the market, almost to the point of threatening the extinction of the plant. In some parts of the Ozarks it was grown commercially. "In 1904 among many other products, 135 pounds of Ginseng was shipped out of Crawford County by railroad," quoted an old copy of the local newspaper now exhibited in the Steelville museum. Most of the shipment was from the wild.

Later we drove through Sligo and down to our farm where we got lost in the woods and had to backtrack to the east end of the farm in order to orient ourselves—lost on our own farm.

On June 17 we camped on Crooked Creek at the base of sandstone bluffs. At the top of the bluff, down low out of reach in fissures between the rocks, a nest of baby buzzards attracted our attention. They were still fuzzy white with big black bills. As we looked into the nest from above, they growled like bears and hissed, but that was not what drove us away.

Plant collecting continued through Reynolds County, along the St. Francis River and river bluffs. Near where we waded across the flooded river a few days before and now only ankle deep, we found Pink-root (*Spigelia marilandica*) which was entirely new to me. Its row of red candles flaring at the top, edged with yellow all arranged upright in a row along the top of a curving flower stalk was quite out of the ordinary.

We continued through Fredericktown and West Plaines and on to Taney County where we camped on Cedar Creek near the Arkansas line. All presses were filled, so the dried plants were removed and packed for mailing to give room for more and more. Plant collecting went on faster than plant drying could keep up, so it became a necessity to dry blotters at every opportunity.

On Sunday June 22 we explored around Ponce de Leon in Stone County along the James River. The night was spent on Jordon Creek in Dade County south of Everton. Monday, at lunch and drier stop, X.Y. was spreading the driers out in an old abandoned field. From the house across the field a man came near and X.Y. said, "I suppose you wonder what I am doing."

The man answered, "You're welcome, my friend. You are my friend?"

X.Y. said, "Yes, thank you, I am."

"Thank you. You are a real man. Thank you. Are you a scout?" He had noticed the long digging knife which X.Y. wore strapped to his belt.

X.Y. replied, "No, I work for a museum."

The man then said, "Oh, that's wonderful. You can strap a gun on your belt if you want to. I have two at the house. You're my friend. I like you a lot. This land is mine. You can use it all you want to. No one can put you off. How old are you?" he asked suddenly.

X.Y. replied, "Forty-two."

"You look young for that, son. Wish I was as young as you, brother. How old do you think I am?"

"About 35," X.Y. guessed.

"Glad to hear that. I'm 48. If I had water, you and your wife could go down and help yourself, but I have to carry my water. I'm a poor man, you see, son, but everything I have is yours. I'm a friend of every man. I planted

Lespedeza here but didn't get a stand. Just stay here as long as you care. I'm your friend, brother.''

X.Y. who had continued spreading blotters answered, ''Thank you.'' Having finished spreading the last blotter he began to pick up those spread before lunch.

''You have a nice car there,'' the man said. ''I'll be going back.'' He abruptly started back across the field looking backward as he went.

X.Y. looked at me and asked, ''Was he crazy or does he think I am?''

''You, I guess. What else could a man think, when you with a big knife strapped to your belt, ran around throwing blotters all over his field, then immediately beginning to pick up an armful. I think he was afraid of you.''

From there we headed straight north up to Independence then straight east on Highway 36 through Hannibal toward home after two weeks and three days in the field.

CHAPTER VIII

JUNE

Household Duties in the Field

When on a collecting trip for more plant records to build up the "Flora" one obligation bestowed upon me in the field was that of tearing newspapers. Since printers never intended their papers to contribute to botany, it was necessary to tear the sheets apart at the fold. The size of one sheet similar to that of a "Chicago Tribune" was near the right size, 12 by 14 inches when folded to enclose a specimen of the necessary size for a herbarium sheet.

After supper dishes were put away, I would tear and fold hundreds of sheets for the next day. That could be done by moonlight, the light of a camp fire, or by the dim outer edges of X.Y.'s lantern.

A woman always has household chores at home or in the field. When out for several days or a few weeks the chores go along. A mending kit with needles, thread, buttons, patches, and scissors went as regularly as the medicine kit.

A one and one-half inch triangular tear in a shirt was often the reason why I got behind on the trail. Ever present, barb wire fences capable of keeping out or confining cattle, never hindered us from going over, under, or between the wires. If X.Y. became snagged by a barb, he used brute strength instead of time to get loose, and went merrily on, while I, caught by a barb above or below, or both, or teetered on top of an unstable fence post, took time to unhitch myself, arriving one minute late, but whole.

X.Y.'s theory was that khaki shirts and trousers were the only suitable materials for the woods. My clothes were always of light weight, and sometimes skimpy—perhaps shorts in hot weather. I held to the theory that the Indians practice: chiggers and seed ticks and such insects that take lodging in the skin where the bands of clothing afford them protection are brushed off by the same kinds of bushes and weeds that deposited them in the first place before they had time to dig in. Since they produced the more lasting and more violent results than the flies and mosquitoes, I endured the temporary annoyance of the mosquitoes. Besides there was no knee binding on hot climbs, therefore less fatigue. At one time though, the brief attire was a disadvantage. Going through a woods I stepped on a large rotten log which disintegrated into a mushy mass. Instantly I was stabbed in the calf of the leg and started to run. Looking back to see if a snake might be following, I discovered that the "snake" had wings. I had crashed in on a house of hornets.

There was seldom enough light for mending at night, therefore that was a job for a rainy day, or if need be, I could skip a hike. In spite of strong, new khaki shirts and trousers X.Y. would come in looking as if he had at that moment returned from the last stand of a losing battle—half a trouser cuff gone, triangular tears in shirt or trousers, or perhaps all covered with clay or gumbo as the locality yielded. As for socks, since two pair a day were the minimum, even a big supply to start with on a trip of two or three weeks would eventually end in more domestic duties.

Had detailed description started soon enough, a long paper or a short book might have been written on Missouri gates. No expert draftsman designing the gates of commerce could equal the ingenuity and variety in gates devised by farmers with materials at hand. A gate commonly encountered, and perhaps easily enough managed by the inventor robust enough to keep it taut, was the two pole, upright, four strand barbwire gate (the fatigue relieved gate). When I opened one of these collapsable contraptions, the wires tired of being stretched, relaxed into a tangle of mulish materials that required time and patience to undo. It was always a relief to find one with an upright pole through the middle. At least the tangles were more manageable.

Another problem gate was the one made of heavy, green oak lumber that had to be pulled uphill to open (the opposing forces gate). Its persistence to go shut in spite of me demanded that I hold on to a bush with one arm and the gate with the other until X.Y. drove through. Then I could turn loose of the bush and run down with the gate, otherwise it might smash to pieces if I turned loose. Its opponent of the same heavy material, opening on an immediate downward incline, had to be lifted or dragged back to clear the way, which was a task for a man.

A third one suggested by nature (the post, bar, lazy gate) was easier to manage. Two trees growing side by side less than a foot apart, perhaps suggested to the gate builder that bars would require the least effort. A few slats nailed between the two trees, and four or five saplings, cut and trimmed for bars to slide across the road and, presto, a gate. There were endless varieties of gates to say nothing of their concomitant fasteners.

I was always the official gate opener and closer unless the gate was beyond my physical strength. Sometimes it was all both of us could do to enduce the gate to meet its fastener again. Given time, their secret or mysterious fasteners could be solved. Unless we had the car with us, we never bothered about opening gates.

We stopped at Pershing State Park in Linn County on June 25 where an interesting drink was made by the keeper of the park. A quart of boiling water with two tablespoonfuls of molasses, one and one-half tablespoonfuls of vinegar, and a pinch of ginger taken hot or iced according to choice, became a good energy recharger. After sipping the delicious drink and a rest in the shade, we visited areas in the park listing 35 species of plants.

It was our custom to take the weather as it came without shelter except for the car, and make the best of it. However, on June 27 when rain was expected, we took a cabin at La Platte in Macon County for two reasons. X.Y. had developed a bad case of poison ivy, and added to that, a second poison from overuse of the remedy he was using for poison ivy. He needed hot water for treatment. He had been pursuing the fallacy that if a little is good, more is better, to a point of danger. Besides badly irritated skin, his feet and legs were swelling. Fortunately after removing the repeated and accumulated doses of calamine lotion on which he had insisted, he felt better. A hot soapy bath followed by a good night of relaxation revived him for plant hunting the following day.

Later at Huckleberry Creek where we dried blotters and ate lunch, X.Y. searched for Huckleberries finding only a few where many used to grow.

Near Tarkio in Atchison County we witnessed a transformation of what had been a shallow lake. Until the year before, lotus and other aquatic plants had grown there. After the lake was drained, plowed, and planted with crops only small patches of bottom prairie remained to give a clue of what species formerly existed. Most of the aquatic vegetation had been exterminated and the present picture showed that the drier types of prairie species were invading and gradually taking over. For example, the *Bidens involucrata* was forming chocking masses over the aquatic *Sparganium*; and *Apocynum cannabinum* and *Teucrium,* more or less weedy species were invading the original stands of *Iris, Scirpus,* and such swampy plants. Such derangement of long-established biotic plant communities brings about an imbalance in relationships in both plant and animal populations for that area. We have pests, plant, insect, and mammal when an existing natural balance is for any reason upset.

On the last day of June we headed homeward. On every trip X.Y. vowed he would leave Missouri in time to reach home at a reasonable hour. Almost every time he failed. The temptation to work in Missouri always held him until late. This time, early in the morning he surprised me by insisting he would cross the Mississippi River bridge at St. Louis no later than three in the afternoon. I bet him a chocolate shake against a banana split that he would not make it. We crossed the bridge at dark and arrived home at two-thirty in the morning with time out for a banana split for me and a chocolate malt for X.Y.

A few years later on June 20 we came into Missouri through Hannibal to do some plant research along the Chariton River, where we took a four and one-half hour hike. In the woods near Shoal Creek we heard a commotion in a big tree at our right and looked up in time to see a huge groundhog-woodchuck frantically clutching at small twigs too frail to hold his weight. He fell to the ground from a distance we guessed to be 70 feet since he was near the top of the tree. The trunk of the large tree had a slope of about ten degrees from the perpendicular below the first branches which enabled him to climb.

Evidently he had gone up for acorns and ventured out too far. Blood was coming from his mouth, however after the first stunned moment he recovered enough to run the short distance to the creek where he died. In the life of mammals there are those that rise above the norm for their species. This one had done so and paid the price. That was one less groundhog to plague the farmer.

On the road to Brookfield in Linn County we picked up a hitchhiker who told us, "If you take and go down that road you'll run into high water."

He saw we were from Chicago and asked if Chicago was larger than St. Louis.

X.Y. pointed out the New Jersey Tea and asked him if he knew it.

"We call that life tea," he answered and added, "My aunt had t.b. and drunk life tea and got well and fat. Now she's really fat."

He kept up a rapid conversation in a voice that jumped from low C to high C several times in one sentence. He informed us, "My brother has a trained rattlesnake. He can make it rattle like a Model T with both ears down. He can make it rattle twice, three times or any number."

X.Y. suggested, "Maybe your brother held up two fingers when he wanted two rattles."

"Or maybe he, the snake that is, could turn off all but two rattles," he continued. Among the traits he might have lacked, imagination was not one of them.

He said, "I know an *old* man about 30 to 38 years old in Chicago. Used to have some relatives there."

When he reached the road where he wanted to go, he left the car saying that his Ma would boot him if he didn't hurry back. Neither was inhibition one of his traits.

Again when we left Missouri at the Chain of Rocks Bridge it was six in the evening and the Old Man River, really the king of waters, was almost out of the banks on the St. Louis side.

We went to plant hunt on June first, four years later near Osceola on the

Osage River. As hot as it was, we had to build a fire because our Coleman stove had failed us.

The next day, for the second time on a Missouri expedition, there was a rainbow in the morning which according to past experience might bring trouble. Disregarding any thought of rain we left the car in a field at the bridge over Niangua River, and started directly up the steep hill on the same side of the river as was the car. Near the top of the hill we separated, X.Y. going west and I going east.

We walked for three and one-half hours. Since it was planned as a long trip I did not hurry. I took time to search over one hill after another, up and down one ravine after another, and I kept bearing more to the south than I realized. I was carrying specimens of a new fern and False Dandelion new to the county, and notes of a Showy Orchid also new to the county. The day was hot and not the faintest breeze ever found its way to the ravines or slopes. About the middle of the afternoon I went down the big hill which I thought went down to our road. It did lead to a road, but to my surprise on the wrong side of the river. I knew I had not crossed the river.

By this time I was hot and tired and now confused. Sitting on a mossy bank of a roadside I tried to locate myself, all the time hoping a car would come by where I might get some directions. None came. How could the river be in that place without my having crossed it? I had to go somewhere, and could not possibly retrace my steps. I had been over too much area. Going down the hill seemed the most logical thing to do. Soon I came to a Y in the road and realized I had come out on a different highway. Now which branch of the Y should I take, or should I go back up the hill and go the other way? As I recalled the ravines I had examined, and ascended to the ridge where the ravine originated, each ridge had been perceptibly bearing to the south instead of running directly east and west, therefore it must be reasonable to go down the hill on the left fork of the Y.

I did make the right choice. It was a great relief to see the bridge and the car even though now, after the confusion, the car was on the wrong side of the highway on the wrong side of the bridge. What tricks the mind can play when one gets lost and comes upon a familiar object. The map showed that the river which had confused me was a winding tributary that touched the highway where I emerged from the woods, placing me south of the river whereas I had started north of the river.

X.Y. was jubilant on his return, for he had found the yellow Barren Strawberry (*Waldesteinia fragarioides*) for which he had hunted 15 years. It is found in only three counties in Missouri. It was a profitable trip after all.

The morning rainbow had a delayed action, but a violent one in the night. It not only drove us into the car, but also out of the river bottom before the river could rise.

After a cold breakfast in the car, X.Y. did an odd thing. He put on all clean clothes throughout and started hiking through the woods in heavy rain. For three hours he wandered and slushed through wet woods in the rain. No day was ever too hot, too wet, too dry or too cold; or distance too great or place too isolated to halt a planned hike. It may mean nothing more than negative evidence, but that was important too.

That night at Dry Fork, since the gas stove would not work, X.Y. chopped up an old stump and we had fire for drying ourselves, and cooking a good supper though nothing could be done about wet beds. It was midnight when X.Y. had finished putting the last specimen into press.

Sitting flat or kneeling alternately by the light of a candle, camp fire or lantern, he persisted from plant to plant. He started with a numbered, folded sheet of newspaper; then arranged a plant between the folds to present both sides of a leaf and various parts of a flower; and then entered the same number in a notebook with all pertinent facts added. These facts such as: date, place, type of location, color of flowers, and any other characteristic that might be lost in drying, took up much of the time.

If he came in with a fat press after a hard day of hiking, this could become the most trying part of the day, and never postponed except on rarest occasions. Sometimes the persistence under extreme fatigue was beyond my understanding.

Fire from the old stump was still glowing by morning, and that was a signal to make hotcakes for breakfast. Later in the day at Elk Prairie we swished through a wet clover field to examine the prairie and found Ragged Orchids (*Habenaria lacera*). There was some concern about this rare virgin prairie becoming a man-made lake in the future.

Upon leaving there we searched for a natural prairie around Jake Prairie, though found none except a tiny corner of the old cemetery.

On the road to Mark Twain Park we bought fresh strawberries from a farmer for 15 cents a quart, and a chicken at Rolla. I stemmed the berries on the road as we drove. It was nine in the evening when we arrived at the park, and eleven o'clock when the chicken and potatoes were cooked.

That night it turned cold for June. X.Y. said it felt near freezing. The radio announced 47 degrees and a man at the filling station at New London said his thermometer said near freezing.

From a man along the way near Sigsby, we learned that he used Buttonbush (*Cephalanthus occidentalis*) for kidney trouble in horses and people. He said, "Boil the roots in an old iron teakettle, or any teakettle. It won't hurt it." It is an old Indian remedy.

We took a hike through wooded ravines bordering South Fabius River where X.Y. got Maidenhair and Christmas Ferns, and Grove Sandwort (*Arenaria lateriflora*), I found the double pink Bindweed (*Convolvulus japonicus*) homeward bound. This time we managed to cross the Mississippi River early enough to have dinner in Springfield, Illinois.

CHAPTER IX

JULY

Ghosts Left by Floods and Progress

It was a 300 mile drive from our home in Illinois to the nearest entrance to the state of Missouri. Add to that our first point of interest, the presently flooded Missouri River anywhere from ten to 250 miles, and our nonstop destination was fixed. Departure time from home in Illinois might be any convenient hour, from three in the morning to six-thirty in the evening. This time, on July 19, we left at seven-thirty in the morning with the prospect of a heavy rain coming up.

We entered Missouri at Hannibal, and drove westward over Highway 36 clear across the state to St. Joseph, then turned north to Big Lake State Park in Holt County for camp at ten that evening. The persistence of a man dedicated to a project was excelled only by the recently flooded river surging to its outlet, surmounting all obstacles. After fourteen and one-half hours of driving X.Y. stopped for the night while the river roared on.

By the time we arrived, bright stars furnished light to guide our moves in the open grassy spot we chose for the night. When settled, X.Y. looked lost, but as snug as a bug ensconced in his sleeping bag in the tall waving grass. The wind blew a gale throughout the night, snatching my covers from the cot, and stirring up or stimulating crickets that made merry by frolicking over our beds all night.

At breakfast the wind blew our cold dry cereal out of the cups and off of our spoons. We thought it wise not to try to cook any breakfast there in the grass and wind.

One chief purpose of this trip was to see what kind of plant life was left after the big floods had swept the banks of the Missouri and Mississippi Rivers and their tributaries. First we drove to Craig facing the loess mounds of Holt County.

The flora of these grassy mounds resembles that of the Great Plains. We climbed straight up some of the hills, a feat more difficult for the cattle. However, their manner of travel over the hills was a better practice in conservation. They went back and forth like highway switchbacks following the contour of the hills rather than straight up and down. These ribbon-like paths woven back and forth across the hills at intervals of ten to thirty feet apart were visible for one-half mile or more.

We found a few things that had escaped grazing such as Thimbleweed

(*Anemone cylindrica*), the rarest species of *Anemone* in Missouri; *Yucca glauca* found only in Holt and the county north of it; the fruits and one lone late-blooming blue flower of a *Penstemon grandiflorus*; and in a slightly shaded ravine between the hills, much Snow-on-the-Mountain (*Euphorbia marginata*), which on this hot day gave the false appearance at a distance of snow in the valley instead of on the mountain.

After wandering up and down those hot, bare hills we drove southward to Mound City. There X.Y. explored more loess mounds while I held my aching head.

From the mounds we passed through the small town of Fortescue near the Missouri River. It was one of the many towns ruined by the spring floods. All day we drove over land which had been flooded. Late in the afternoon X.Y. explored near Forest City, though I sat still, nursing my head in front of what had been a very nice farm house. Now the gate was hanging off to one side, rock walls were sliding down, the attic barn door was slamming in the wind, a buckling garage filled with farm equipment was deteriorating, and a bike now sat useless with a load of mud and trash. Vines, limp shrouds, and pieces of clothing stripped from the now-buried lands hung on the telephone line 15 to 20 feet above the road. Buttons reflecting the setting sun were now beyond the reach of human hands. X.Y. returned with a few scrappy plants dug from the mud and we drove on.

Near Forbes, a small, half-grown red fox crossed in front of the car, ran up the slope, and stopped to look at us. As we approached he ran farther up the slope, went behind a bush and peeked around to watch us as we passed. A boy hitchhiker said, "If you see one fox, you can bet there are a dozen, just like as if you see one rat there are hundreds."

At Oregon we filled the icebox and drove to a creek to camp. While I was getting supper a man and woman and their daughter drove up. The woman, seeing our icebox, Coleman stove, and mosquito tents said, "Now that's the way to camp. Have you had a good catch?"

They went on down the creek to fish, but soon returned with nothing more than mosquitoe bites.

X.Y. also had developed a headache from too much hiking in the hot sun. We both went to bed at dark. This was quite an exception to the customary routine. X.Y. usually put the collection of the day into the press while I got supper or he finished the job after supper. Then he would fill in the county records. He carried a set of notebooks containing small Missouri maps for every probable species of plant in Missouri. Everyday he placed a cross in the county where a species of plant was collected or seen, if the county did not already have a cross. It was always a big day when a new county record was found and a cross added in its proper place. And it was a very special day when a new species could be added to the state record and he could write the name of the new species under a new Missouri map and fill the cross into the county of a perfectly clean map.

On this particular night, it all waited until morning. Ice melted fast in the 105 degree morning that followed. Again we stopped at Oregon for ice and groceries, then headed northward. Near Nishnabotna—the Indian name for crooked river—we stopped on a side road to collect and photograph Blue Lettuce (*Lactuca pulchella*), a large blue-flowered one known only from Atchison, Nodaway, and Jackson Counties. It resembles a large blue aster. I barely pulled the camera and tripod down in time to run from a tractor. People in passing cars kept looking at us standing among the weeds in the boiling sun as if we were queer. Lack of perception and conception are words descriptive of our spectators. The only conditions these people were aware of, were those perceived by the physical senses of sight, sound, and feeling, but mainly feeling on this 105 degree day.

At Nishnabotna we drove over the flooded bottoms of the Missouri River to a dyke and over the dyke to fields of new white sand laid down by the spring floods. While we walked around on this new sand in 105 degree temperature, X.Y. found a yellow *Rorippa* common in the West, and a Tamarisk which had washed in from western states by the flood.

Almost within sight of the dyke over which flood waters had poured, a new house sat on a mound of sandy soil by the side of the place where the road had been. Water had been up into the house. All through this area flood waters had scooped out holes and piled the sand into mounds somewhere else. The owner of this home was elevating the foundation of the house. In view of surrounding results of the past flood, it was difficult to imagine a faith in the future that would not picture the house being moved to the scoop across the road during the next flood. This was their land, and they and others like them, built right back on the same unstable site even though their rich, black loam was covered by feet of pure sand. They were here today and perhaps gone tomorrow determined to hang on as long as possible.

In this section houses were all buckled outward, tumbling down, and at present deserted. Roads, if at all passable were traveled at two miles per hour. At Nishnabotna, flood waters had stood at nine feet in the store and post office.

There was little to find in the way of wildflowers on the farm road to Tarkio and back southward where we found a campsite in a deep woods near an untraveled side road. It was a hot night. A good breeze rustled the tops of trees and drew out some of the hot air. A great horned owl gave a few hoots then quieted down for the night, contrary to the usual all-night concert. It was too hot to hoot.

The next morning as we drove along the road toward Nodaway County after breakfast, dickcissels were singing incessantly on every wire. Yesterday the wires must have scorched their feet. We added another record of *Lactuca* in Nodaway County and later made a stop for *Ampelamus albidus* in Gentry County after stopping for more ice and groceries at Stanberry. We seemed to be collecting as much ice as plants.

At Pershing State Park we had lunch and both worked at changing the driers. At the last place where the driers had been changed we spread the grey felt blotters over an old field. One man stopped and asked if we were trying out a new kind of fertilizer on the field.

Near Rothville we hiked over cattle paths across a hot field and down to a dry creek bed. There we had a treat. Several species of Blackberries were growing in confusion on the banks accessible from the bed of the creek. One vining Blackberry had climbed eight to ten feet on the shrubs of the bank. This one was heavy with lush, large juicy fruits—their juices drawn by roots thrust deep down into the sand of the dry creek bed. Long strands hung with succulent black fruit. We ate until the berries tasted like water. They were the largest and most abundant Blackberries I had ever seen, and did serve to quench the thirst that troubled us at that time.

The layer of yellow laundry soap I had rubbed on to prevent more poison ivy than I had already picked up was becoming intolerable in the heat and I left the Blackberry feast to go to the car and wash off the soap.

July heat speeds up insects. Chirruping crickets call more rapidly, and ants that usually creep, actually run to and from their sources of food supply when the temperature soars. My journey from the Blackberry source across the hot, dry field of sparse wilted grass and a few scattered stunted Rose Gentians was slowed to a stroll that was more aptly called a "drag."

Another object of this trip was to review the flora of salt springs, therefore, we drove to Glasgow on the Missouri River to begin a search for salt springs in Saline County. An old lady rocking on the lawn had ten acres, but no salt springs. The first man we asked had no salt springs and did not know of any, however, we found the large Blue Lick about one-fourth of a mile from where he lived. This onetime salt lick had covered an area about 600 feet wide. Now it was all dry and crusted with bare hard soil except for a swampy salt marsh where Bulrushes and Narrow-leaved Cat-tails grew about 50 to 100 feet below the spring. The main channel from the spring once about 50 feet wide, now carried a stream two to four feet wide filled with blue-green algae. Near the spring, the water issued with a milky-blue appearance. The main part of the spring was enclosed in a two-foot-square wooden box. Also in a small draw nearby, where a foot-wide shallow stream came from under a tree, water bubbled up out of mud. All rocks and soil were salt-covered as far as 100 feet below the spring. A brackish odor was given off over the whole area. The area was only a miniature of its former extent. I stood and tried to reconstruct the picture that might have been seen thousands of years ago. One man unconcerned about the future once said to me, "What my children do not know they will not miss." That attitude taken so lightly by so many people has been the basis of the existing and rapidly approaching crises of the present time. At this time I was wishing there was some way to know the past of this area. Too bad there were no photographs, books, or specimens to support the imaginations stirred by the minute remaining relic. X.Y. was trying to salvage the remaining bits against the time when the entire lick will cease to exist.

After lunch and a change of blotters, we resumed our search for salt springs and vegetation growing in such habitats. At Elk Lick the spring was free of plants in the center where the water appeared to be more saturated with salt and sulphur, but Bulrushes grew in the area 50 to 300 feet outward from the spring.

For sometime X.Y. and I looked for a salt spring where in October of 1938 he had found the only station for *Distichlis spicata*. This had proved to be the only station in the interior of United States ever recorded outside of its coastline stations where it is confined to the saline soils of the Atlantic, Gulf, and Pacific coasts. Here he found this isolated relic habitat of a large prehistoric salt region squeezed down to the small area of a few acres. Our aim at this time was to find any other plants that might be growing with it in the month of July.

X.Y. was remembering the area as it appeared 14 years before. There seemed to be nothing but corn fields on all sides. We ran up and down rows of tall corn, fields of weeds and wheat stubble, finding nothing familiar. Finally we went to a farmhouse. There we found a 13-year-old boy who knew where to find the spring. He took us in the opposite direction from where we were searching. That spot looked quite unfamiliar and no trace of *Distichlis spicata* could be found. Flooding and cultivation had altered the region beyond recognition. Could it be possible that this relic which had persisted for millions of years had been completely wiped out within the last 14 years?

The boy said there were millions of mineral springs in the Blue Lick vicinity. He dug through a three-inch mat of *Eleocharis parvula,* a saline sedge, to reach the clear mineral water which tasted like imaginary over-salted rotten eggs—more sulphur than salt. Duckweed covered every inch of the water and a crust of white deposit spread outward for many feet from the spring. In the photograph I took it looked like snow against tall green corn in the field beyond. Disappointed that the rare plant was there no longer, we left the locality for Marshall and southeast to Black Water Creek for camp. A great horned owl hooted for an hour over X.Y.'s head, and he dreamed of owls until his tent poles fell down.

Thursday morning was a cool 65 degrees, but clear. After a hike over chert-covered bluffs in Saline County, we drove to Marshall for ice and groceries, and on northward to a swampy section, now dried up at Grand Pass near the Missouri River. Later where some men were working on a new bridge, X.Y. stopped to ask the way to Lamine River or to New Lebanon. Again they assumed that we were fishing. They could not understand what we were doing or why we were wanting to find isolated bad roads. They started right away, as did most people we encountered, trying to direct us to some town nearby or of greater size, or to a good highway.

The following day we drove to Boonville for ice again, then to Otterville where I changed driers on the roadside near the bridge over Salt River. I nearly roasted at 96 degrees and no breeze. X.Y. was trying to find *Trillium nivale* on the bluffs nearby where he had formerly found them. He came back with a load of sweet Blackberries which helped to quench the thirst since we had no safe drinking water at this time.

On Spencer Creek south of New London we made camp where music kept me awake. That is it kept me awake in a sense that I deliberately stayed awake to listen to the incessant argument of hundreds of katydids repeating, "Katy did, Katy didn't;" the bullfrog bass "Jug-o-rum" followed by the sudden splash when he flopped into the water to moisten his drying skin; the long-winded whip-poor-will; and the sawing and sizzling of crickets like myriads of tiny electric saws. I finally went to sleep trying to count the number of whip-poor-will calls one bird made without getting his breath. Life on a small swift river playing tunes on rocks, accompanied by its nocturnal denizens, if perceived by the mind tuned to its vibrations, is a most satisfying and relaxing diverson.

It finally cooled off during the night for comfortable sleeping. After breakfast X.Y. went on a short hike, and I walked down the creek and crawled up a steep bluff to a glade where I found Purple Coneflower, Foxglove, and *Phlox paniculata,* a new record for the county. I descended by way of a beautiful ravine with moss-covered banks bordering waterfalls. It was a beautiful place to pass through but only as fast as possible, because I could not outrun the mosquitoes, deerflies, and horseflies. I called it "valley of the flies."

At ten-thirty in the morning we left and drove to a side road off of 61 between New London and Hannibal to eat lunch and change driers. Everyone who passed slowed up to ask what had happened when they saw the blotters strewn about. Fields were now beautifully covered with red clover humming with bumblebees on the sunny hot day of 100 degrees. Hot haze over a red clover field, and a concerted droning of bumblebees busy in that field, following our ample lunch, anesthetized us to sit in the shade and simply float at ease in the abstract—take a trip.

Finally we had to shake this inertia and head north of Canton toward the Mississippi River where we stopped at a ditch as green as grass with dense mats of filamentous blue-green algae. Small fish were dead and floated by the hundreds. Some were still gasping for breath. Whether they were dying from a gas given off from the algae, or from some commercial byproduct or chemical drained into the water, I did not know. Since one species of alga resembling the one in the ditch does give off a gas capable of killing fish, it could well have been the cause in this case. At one time we lost 40 kinds of tropical fish from the gas of algae introduced into the tank. The quantity of gas given off regardless of efforts to control it was even offensive to us in the house.

From this green-covered ditch we walked down the dirt road across Doe Run then stopped for awhile at Buck Run where I counted 33 egrets in the trees over a partially dried up swamp. Under the edge of the bridge where I stood, a water snake with a body as large as my forearm lay within six inches of a little turtle no larger than the snake's body. The snake ducked and came up with his head only three inches from the turtle on the log. After about five minutes the turtle slowly turned its black and green striped head to look at the snake then looked straight ahead unconcerned. At the stamp of my foot on the bridge 15 feet above, the turtle dropped into the water, but the snake never batted an eye. We left while killdeers and sandpipers still bobbed in the mud along the stagnant water. Egrets and great blue herons were common along the road.

After driving to three dead end roads we finally found a place to spend the night by a dried up creek. Poison ivy created a problem here as well as lack of water, however, we decided to make the best of it. X.Y. went west along the sandy-bottomed creek in hunt of water and found none. I went east and found a hole about eight feet across and 18 inches deep with clear water suitable for all purposes but drinking.

Again it was hot and dry, and we drank ice-cold lemonade which became weaker and weaker, until the ice supply ran low. We sponged off with ice water which drained off from the melting ice. X.Y. worked until eleven o'clock putting plants into press and recording county records. He had found *Sabatia angularis,* a new record that far north.

The following morning X.Y. parked the car by a church and went on a short plant hunt. I looked after drying the blotters and studied the trees across the road from the church. Two White Oaks with diameters 15 to 18 inches each formed a perfect arch toward the east. A third one leaned northward. One White Oak and one Hickory each formed an S curve. A smaller Hickory and a small Elm arched westward. It was not the wind that did it. Such a curious-looking bunch of trees in one spot testified not of religion in the church, but of small boys riding saplings outside while church went on inside. As the sapling is bent so the tree grows—at least that went for the trees. As for the boys, who knows.

At ten-thirty that morning we crossed the Mississippi River at Hannibal homeward bound for Barrington after one of the hottest expeditions with one of the greatest consumptions of ice water and ice-cream in any one week of plant hunting.

CHAPTER X

JULY AND AUGUST

Search for an Earth Burp

The precipitations for this year practically all fell at close intervals to produce floods. We knew before leaving our home in Barrington that flood waters had closed the Chain of Rocks Bridge at St. Louis making it necessary to cross the Mississippi River over the Free Bridge. Even there on the St. Louis side, water could be seen over the lower floors of business buildings.

Beyond St. Clair we carried our fourth of July lunch and jug of iced lemonade to a sunny glade in a wooded area quite isolated north of the highway. A lunch stop could always be used for reconnoitering between bites, and often between bites interesting plant records were found. Standing up to eat, of course, was a rest after a 350 mile drive. We had many more miles to go to reach the Niangua River, our aim.

Later, two miles south of Bennett Spring State Park we camped on a gravel bar. Rain drove us out at four o'clock. To be on the safe side of the creek we packed everything and crossed it. While X.Y. took a cold water shave in the dim light of early dawn, I carried things back to the creek to start breakfast. The coffee water had merely begun to steam when I had to pour it out and scramble everything into the car again. We then drove to Bennett Spring State Park and cooked breakfast under shelter while it rained.

The first habitat listed for inspection was a natural pond where we found Meadow Beauty (*Rhexia virginica*) with attractive rose-red flowers. We also found a pink Rose Mallow (*Hibiscus militaris*).

Besides the highway and contour maps which X.Y. carried, he carried section maps for every county. These maps were invaluable for locating swamps, natural ponds, sinks, and canyons, and as a further aid for following side roads, churches and schools were named on these maps. There was even a dot for each house along the way. Sometimes we counted dots to see if we were on the right road.

One preliminary preparation before leaving home was for X.Y. to mark on the section maps, the places he chose to visit. These maps were then filed alphabetically in a fat briefcase to be juggled by me as we proceeded. Often the mere name of one lone country school sitting idle in the weeds indicating that the former familiar feet were now dangling from bus seats enroute to a consolidated school put us again on the right road.

South of Dry Knob along Osage Fork we examined a limestone glade and bluffs, though found little but moss-covered rocks and ferns.

It was getting late as we drove down a badly eroded old road to the river and found that the bridge over the river was washed out. There was nothing to do but make camp there or drive back up the bad road. We made camp. There was not a cloud in the sky, and soon bright stars were lending cheer, and a concert of many frogs ranging through more than an octave were whooping it up. Crickets with a high-pitched friction orchestra added to the nightlife of the low, damp recently flooded river bottom.

X.Y. said, "We'll sleep in peace throughout this night."

I answered "It will rain about midnight," and really believed it. The air felt like rain regardless of the starry sky. A low barometer (low air pressure) brought forth musty smells from the swampy, recently flooded river banks. Gases formed by decomposition of vegetation arose from grass and weeds along the banks with scarcely a breath of air to remove them.

"You're a nice pessimist," he informed me.

All went well until one o'clock when lightning began to flash and thunder began to roll from cloud to cloud. I got up and began to dress. We were not in a safe location for rain. The river was too close, and the banks too shallow for any more water. I was dressed and had my bed almost packed when X.Y. grunted from his sleeping bag, "I hear thunder. Get up."

We barely got things packed and had just pulled out to drive to the top of the hill when the storm broke. By three o'clock the rain had stopped. We got out on top of the hill and made our beds again where we rested about an hour. Rain again drove us into the car and we drove on to a small creek. By that time the rain had ceased again. We piled out and were scarcely settled when rain again started. I loaded my bed and vowed I would stay in the car. X.Y. covered by a rain coat, just let it rain. He was close to the little creek that was rising, and every few minutes I would look out by flashes of lightning to see if the water had reached him. There were so many rain-filled puddles all around him it was sometimes difficult to be sure if he was in the creek or still on the banks. What a way to spend the night. Neither water, heat, insects, fatigue nor sleepless nights would show through a plant in the "Flora." Some dogs had come along before daylight and were either curious or alarmed. One small black and white dog sitting in the road across the creek started a shrill, "Yip, yip." Then a lanky hound joined in with a husky baying, followed by the yelps of a small black dog. Two more joined the chorus and our privacy was ended.

Everything was wet. A cold breakfast had to suffice.

Bushes were beginning to dry somewhat when we left the car at the top of a ridge and walked down a washed out winding road one and one-half miles to Pine Creek where beautiful high rocky bluffs rose from each side. Strangely the creek was clear, though not so strange when the forest floor was considered. Here there was heavy leaf mulch and dense vegetation to hold back the soil and filter the water.

X.Y. had gone out without a cap. Deerflies, the small triangular pests somewhat smaller than a housefly, light on the head or some other exposed place and hang on until they draw blood and leave a poison worse than that of a mosquito. He devised a scheme to defeat the flies that were biting right through the red bandanna handkerchief he had on his head. By tying the bandanna around his head in the form of a band and sticking oak branches under the band he found that the waving motion of the oak twigs kept the deerflies away as he walked.

It worked very well until we came out on the ridge road. I was taking pictures of his headdress when some people going to church came by and ran into a mud hole. Ridge top roads through forests always seem to have a few mud holes. Though high and seemingly well drained, for some reason the fine gray silt road bed gives way in spots that have no bottoms. The farmer or regular traveler who uses them simply cuts the small trees by the roadside to bypass the hole, and if that gets a hole they bypass on the other side. It is not uncommon for a stranger who comes along to find three roads side by side, each with a sink hole—possibly filled with brush—and wonder which of the three is the lesser evil. At this time the man may have been distracted by the queer-looking shoofly device X.Y. was wearing and carelessly landed in one of the depressions. Now X.Y. had to discard his headdress and try to help the man get out of the mud.

On the way home on Highway 66 we took pictures of Devil's Elbow though the Kodachrome lost part of its charm because the river was swollen and muddy. The name "Devil's Elbow" was given to the high bluff at the bend of Big Piney River by lumberjacks who feared and cursed the log jams that always formed at the bend.

The devil enjoyed distinction in Missouri. There is Devil's Ice Box, a cavern near Columbia where cold drafts come from an open roof of a cave in the summertime. A sharp ridge along the top of a hill acquired the name of Devil's Backbone, which also has a rocky slide down a hillside. There is also a very rocky and wild, but clear creek called Devil's Run near an old swinging bridge.

Devil's Elbow is listed as one of the seven beauty spots in the state.

This time we crossed the Mississippi River at St. Louis at seven that night and arrived home at three-thirty in the morning. Another bet lost.

For once on our many trips we left home in Barrington at the reasonable hour of six in the morning to make a hot, summer plant expedition.

At the university cottage where we arrived at two o'clock on this July trip, the heat was almost too much to bear until the windows could be opened. Since no one had used the cottage for some time, they were all stuck fast requiring time and effort to open.

The caretaker, after giving us the key to the cottage, returned to offer us some vegetables. I had already seen the garden and would have refused the vegetables even if we had none with us. This was an extremely dry year in Missouri, and his garden showed it. The reverse was true in Illinois. We had a perfect year for gardens, and from our little garden I had brought along all we could use. Had I known the plight of his garden I would have brought some of the excess from our private garden in Illinois. I had been hauling vegetables by the cart-load down to the house, wondering what to do with them. It was great to have our own unsprayed vegetables straight from the garden, fun to watch the new green rows appear in the spring, and a sense of achievement to pick them even if I did have to call up the neighbors to take them off my hands. No credit need to be claimed by us however, for the very next year with the same seeds and hothouse plants, and the same care and no rain, I carried in two hands what I had needed a cart for the year before. We told the farmer caretaker we were well supplied.

Having opened the windows and established ourselves with no fear of rain, we drove to our farm in Dent and Crawford Counties. There we found more water in the creek than expected considering all the dry weather. We entered at the lower, north end through our neighbor G's farm, and walked across his pasture to where the four majestic chestnut trees had been growing for many years in a row on a ridge. We had passed there a few years previously when chestnuts were ripe. Now one big tree was dead (in four more years they were all dead). For one hour we talked to Mr. G. who was harvesting hay.

The day was hot and we found it pleasant to wade up our clear, gravel and rock-bottomed creek all the way to the far east end. There across the fence the creek bed continued without a creek. Bare dry rocks minus a trickle of water glared a menancing threat to me across the fence. The man who owned the large section of land above ours had overpastured the wooded headwaters clear down to the rocks, and the spring feeding the creek had dried up. Not one green plant could be seen within the reach of cattle. While such a barren site above our farm did affect our creek, especially for a few yards, the dense growth and water-holding mulch of our own unmolested creek bottom could still feed the creek enlarging it as it flowed farther on. We wanted a drink from our spring, but had only leaves for cups.

Our springs were still running, and the creek had some nice swimming holes where X.Y. had a swim. Then when he went on farther to look for plants, I sat on a shaded gravel bar and relaxed to experience the contentment that goes with a piece of land owned which pays for itself in aesthetic measures only: the tinkling of a clear brook; cool mossy, fern-covered bluffs across the creek rising from the water, though not large, charming nevertheless; birds flitting in overhanging branches of trees from which an occasional bird darted down to a shallow spot for a bath; and the quiet privacy where anxieties and burdens dropped away, setting me free from the impedimenta of life. I sat there cool and dreaming, and collected various forms and colors of pebbles about me while X.Y. worked.

He returned with a press full of plants from our farm, and I took a pocketful of pebbles. We walked back to our car on the road and drove back to the cottage, very hot, tired, hungry, and sleepy since we had left Chicago so early in the morning.

The cottage was still too hot. I cooked on the outdoor fireplace, and we ate on the porch. Later, we finally carried our camp beds to the porch and slept out to keep cool. We felt like drinking the spring dry of its cold, clear water.

After a late breakfast the following morning we set out to find the pinnacles on Crooked Creek. We thought the best approach to the pinnacles might be through Steelville, then out west to Highway 8, and up the valley through Keysville.

When it appeared that we might be near the vicinity of the pinnacles we stopped at a farmhouse to ask, but the owner never had heard of them. We started to hunt for them—hiking for one and one-half hours, up and downhill, and over valleys. Some people fishing had read of them, though they had not lived there very long. They told us that a farmer one-fourth mile from where our car was parked had lived there 30 years and he might know.

The number of people who are mentally blind and deaf to sights and sounds surrounding them, far exceed the number of people physically blind and deaf. One sees what one wants to see, or more accurately what one has been trained to see. The trained botanist may see only plants; the trained architect may see interesting architecture at every turn; and the mule trained to the rough trail sees only the easiest pass, or obstacles to be avoided. To the botanist, the architect, and the mule all else is only a blur along the way.

The two literary characters—rich Kansas widows, untrained in history or the appreciation of historic places, went around the world and saw only the homely things that reminded them of their lifetime Kansas homes. A woman, when asked about the song of an interesting bird, replied that she had never heard it, though the sound waves from its persistent, exhilarating song had vibrated her eardrums for many years.

We called on the farmer. Yes, he knew where the pinnacles were, for one of the upheavals was in his barn lot. He showed us the scarred rocks jutting out six inches in a series of steps. He told us where to find the elusive pinnacles.

As it was lunchtime we sat on the banks of Crooked Creek to eat lunch and threw bread to the fish, wondering if some of the fish we were feeding had come downstream from the tributary creek on our farms.

After lunch we followed the creek up where it had cut through the cone slicing off a small chunk of the outer part. In fairly recent times geologists discovered that the pinnacle-like formations at the top of the hill were not merely rocks as on other hills, but the top of a cone-shaped area pushed up by a subterranean explosion, a bubble that did not have force enough to burst open. Further surveys showed the cone to be formed of concentric rings of the seven various strata of rocks and soil of the surrounding country, sometimes pushed up 1,000 feet above their original level. The cone is from three to four miles in diameter at the base.

We climbed to the top—pulled up or crawled up over some of the steeper places. The day was unusually hot. While the earth burp elevated a problem to the geologist, who still is not certain of its origin, it caused no problem to the botanist other than that of climbing the hill after it was found. X.Y. found no disturbance in the flora.

We drove back to Cherryville to purchase cakes and lemons, then on to Dillard to look at Simm's Flat, the exposed, long barren ridge. On this afternoon of 100 degrees I tried to bring down the temperature by recalling the February years ago when I had driven over the wide open area in the full blast of a zero snow blizzard without any heat in the vehicle and nearly froze. We found a few Dewberries and some very sweet Blackberries, though almost everything was dried up.

Hoping the fern dale of which someone had told X.Y., would prove more rewarding we drove to Davisville and on to Y's dairy farm. He went with us up the hollow where he had found many ferns. There we found Christmas, Maidenhair, and Rattlesnake ferns.

The copperhead snake coming down the hill toward us had to be permitted to go on its way undisturbed because of the dog. The dog would have attacked it if permitted and likely would have been bitten.

On returning from the breezeless, torrid hollow we stopped at the cave for a cool drink of spring water. The cool entrance to the cave by sudden contrast sent shivers up and down the spine.

We took a shortcut back to the cottage, and after some trials and errors, arrived there at eight-thirty. After supper I found the bathroom too hot for a bath so we went for a moonlight swim in the clear water of the Huzzah River. There we could lie and watch the highlights on moonlit waves as the water rippled over us.

We were searching now, August 1, for a sandy canyon near Elsberry in Lincoln County. We hiked down one narrow gorge, stopping to eat luscious Blackberries one inch long from a vining species which tried to reach the sun for its foothold in the gorge. This was not the canyon we were searching for.

Next we crossed a hot field where Rose Gentians (*Sabatia angularis*) were blooming. This fragrant Gentian, a biennial, likes its freedom from competition with other plants. The sandy or poor worn-out soil of old abandoned fields is not what is favored by the plant, but the absence of other plants that might crowd or shade it. It was found to produce larger and richer-looking plants and flowers when fed fertilizer, and given plenty of open space, but soon failed to reseed or develop when grass and other plants were allowed to compete.

When we see a plant growing in a certain habitat, not all meets the eye. The tendency is to think that the existing habitat is the one most favored by the plant. Our experience with the Rose Gentian revealed some things about the Gentian that we should have guessed earlier. No Rose Gentians grow naturally in Barrington. The nearest were about 60 miles away. I wanted to try some. One late summer day we found many plants in flower and good seed at the Indiana Dunes in pure sand. I took some seeds, and a sack of sand. Our prairie wild flower garden in Barrington did not seem a likely place, however I selected an eight foot square in full sun, scooped out rows about one foot apart, put in sand from the dunes, then the seeds, then a bit more sand, then covered lightly with black dirt. Since the Rose Gentian is a biennial I was surprised that they germinated and made the rosettes that same fall. Next year they bloomed, and did well. The following year, to my surprise, they had spread without any help from me for about eight feet from the original bed. The next year they went farther, and continued until they had spread about 100 feet from the original sand-prepared plot. Then since other things were left to spread at will it soon became evident that they were diminishing. Their main requirement was not soil, but no competition.

At 100 degrees temperature we were still hunting for a particular canyon. Where we stopped for a drink at a farmhouse the woman directed us to a "valley" as she called it. It was a nice canyon where we followed along the base and lower slopes, where not too steep.

When I could find a place to climb to the top, I turned back as usual in order to reach the car ahead of X.Y. Crossing a flat, old field, abandoned years before, I almost stumbled upon a very narrow canyon, like a crack in the earth, narrow enough to step across. The perpendicular rock walls went straight down what looked to be 50 to 75 feet. I had come upon its very origin. It looked as if the rock had split open there. Since it was not possible to descend anywhere in sight, I continued on back to the car.

Near dark when X.Y. returned he gave up exhausted. The day had been too much for him and he went to bed after taking two aspirins and a cup of water. I ate supper alone and was asleep soon after dark.

At five the next morning X.Y. was ready to eat a breakfast of eggs, toast, fruit, and coffee. He went back with me to the old field to look at the narrow canyon, at the base of which we found mosses and ferns where the sun never shines. The Walking Fern had plenty of moisture and no disturbance from drying winds. The narrow steep-walled canyon unguarded and unannounced by any hint of its presence haunted us all the way back to our home in Barrington that day. Even wild animals could be taken by surprise there.

CHAPTER XI

SEPTEMBER

Hunt for Shangri-la Canyon and a Natural Bridge

In this particular year I made nine trips with X.Y. back to Missouri, merely a few as compared with the 20 trips on which he had burned gas and tires in one year during the flowering season, running up a minimum of 12,000 miles only to the border of the state and back to say nothing of the highways and byways within the state. All for the sake of the "Flora."

When we reached Rolla on this trip on September 20, we were not fully prepared. While X.Y. shopped for films for the camera, I drove around the block three times. School was out and there was no place to park. We were then in the high winds and outer edges of the hurricane which had hit Florida and New Orleans, and had given Springfield a sample.

At noon the next day, when we stopped at Ava in Douglas County for lunch, a pert waitress suggested that we eat anchovies. She said she liked them, except that her family was too large for her to afford them. X.Y. sympathized by saying that he liked them too, but it was the same with him. What she did not know was that he was the only one in his premarital family, and the only one of two in his married state.

I thought he might be returning a favor of a former waitress where we had a late breakfast in southern Georgia. For some odd reason he ordered chili for breakfast—it was late and it was one of his favorite dishes. When the waitress brought it, he asked for a hot sauce to add to it.

As we began to joke him the waitress took his part with, "Yes, sah, I go get de sas. Ah allays has hot sas on mah breakfas chili."

We went to Noblett Creek to camp. When we asked a boy by the side of the road how to reach the creek he said, "One and one-half miles, turn right, another one and one-half miles, turn right." Each of the one and one-half miles proved to be three miles. We arrived a little before dark and found a spring with convenient flat rocks, though I had to hunt for wood in the dark to cook the chicken, and the potatoes and vegetables, the later brought from our garden in Barrington.

A heavy fog rolled up as we ate. In spite of the plastic covers with which we tried to cover our beds, they were wet and clammy by eleven o'clock when our work was finished, and the fog-dimmed moon had already gone down behind the hill.

We were sleeping well when the alarm clock went off at 5:30 daylight savings time. Heavy fog still hid whatever prospects of daylight there was. Fog dripped from leaves and twigs and splashed on the plastic. Half of my pillow, where exposed, was wet. We gave up the idea of mopping up the fog-drenched vegetation and returned to oblivion for two more hours.

When the sun began to glimmer through the milky air exposing the tops of hills across the valley, we soon had a good fire from the wood I had wrapped in newspapers and placed under the car the evening before. A hot breakfast took the chill out of the 45 degree morning. Nothing smells better than a camp fire of hickory wood on a cold morning unless it is the aroma of the coffee or the bacon.

By nine we were ready to start in search of a hidden canyon which X.Y. had long wanted to explore. The only road to this Shangri-la canyon was by water or foot. We waded down Noblett Creek because the way was too rough going on the banks, and partly because it would be easier to see where Spring Creek (which drained the canyon) flowed into Noblett Creek. So many steep bluffs along the banks at our left where the creek was to enter prevented our walking on that side. Beautiful bluffs with pendent Maidenhair Ferns almost touching the water and other plants of interest caused frequent halts to collect or make notes.

The clear, cold water rambled over gravel beds the entire length of the stream which we continued to wade in, often knee-deep. It was one o'clock when we arrived at the first house, and we still had not found the tributary Spring Creek. We had been walking steadily since nine with only a five minute stop to eat a fig bar and four crackers each, and to pull up our socks which had worked down into balls underfoot.

The woman at the house said we had passed the junction of the two streams some distance back. It was hidden by a small tree-covered island between us and the confluence. She said they had not lived there long. She had come from Texas where she could see out, and she did not like it here, because the only way she could see out was straight up. They were really shut in by hills on all sides. The only road out was up a very steep hill. She gave us a drink of water and a few small seedling peaches from a tree on the lawn.

We started back down the creek and this time found Spring Creek, which we followed up for about one and one-half miles to the narrow steep Shangri-la canyon wrapped in almost perpetual gloom with perpendicular bluffs on each side. One boulder as large as a house had fallen a day or so before, breaking trees down, smashing a lovely bed of ferns at the margin of the spring branch.

At 3:30 in the afternoon the sun was cutting a rough diagonal across the mossy bluffs rising straight up on the eastern side of the Shangri-la canyon. The west bluffs now in shadow were darkening. On the sunny side contrasting sun and shadow brought out the outlines of ferns and mosses, and bits of color from inaccessible flowers covering the face of the bluff.

While X.Y. grabbed plants as fast as possible from the floor of the canyon, the diagonal line dividing sun and shadow climbed higher toward the rim. We would not be there to see the gloom come over the canyon at dusk after the sun had climbed out. I suggested to X.Y. that we come back if the moon came up to spend the night in Shangri-la canyon to experience the magic which I felt would be unique. I am sure no one had ever done it. Also I would like to have seen the sun come down the west walls of the canyon at sunrise to push the shadow up the opposite wall. There was no time to dream or linger.

I was gripped by the same urge which years before took me down the Black Canyon of the Gunnison River in Colorado (one of my foremost trophies).

Early in the morning with the smell of melting snow mingled with the fragrant wood smoke from the stove in our rustic cabin, giving at the same time a feeling of chill and warmth, I slipped away to explore only a bit of the dark canyon. It wound around with high perpendicular walls, and the river, now swollen by melting snow, followed the curves. A singing canyon wren also followed the curves. I followed the canyon wren. Impelled by a common compulsion we all went down the canyon. The river went because there was no other way for it to go; the wren because, who knows; and I because of an irresistible curiosity of what might appear around the next bend. The hypnotic sound of the wren song and the swollen river pulled me on and on with sometimes barely space enough to walk between the river and the canyon walls. There seemed to be no turn to discourage my advance except the growing consciousness that the cabin at the first turn with hotcakes and coffee was becoming my stomach's vital concern.

These fleeting trophies remain fixed only in a lifetime. They can be captured only once, never to be seen or experienced again in the same manner.

Now we were both loaded with plants. X.Y. was in a hurry, and I, practically running through the bushes to keep up with him, hooked my foot in a vine and fell flat. On the return we were following a cow path which kept us out of water part of the time.

At five-thirty, our campsite came into view. X.Y. made a dash for the icebox and food basket, while I was more concerned about dry socks and shoes. My feet, in water from nine in the morning to five-thirty in the evening, were wrinkled like the small, desiccated seedling peaches we saw on the ground at the only house seen all day. The hidden canyon had given up its plant secrets and had been duly admired.

A hot supper was quickly put together, and our water-soaked feet were tucked in by eight o'clock. Fog came again as on the previous night and we were somewhat chilly as we slept.

At six the next morning after a hotcake breakfast we folded up wet beds and cleared out. We were enroute to look at Buffalo Bog near Cassville. We found the bog dry. The owner said it went dry every year for 30 years, and continued to get drier each year, though it used to hold water the year round.

Following a dry, rocky creek bed down between two bluffs we came upon a tall, thin man with four dogs driving ten cows. We asked him where we could find water for camp.

He told us, "You can camp in that old house there by the road. An old bachelor lived there. He died in January. Found him sittin' in the chair dead. Been expectin' him to die for months."

We thanked him and he went on to assure us, "There are no ghosts in these parts."

He went on with his dogs and cows. Through the broken window we could see the old rocking chair just where the bachelor had left it.

Not far from there we found a place to camp on a gravel bar. I made hunter's stew while X.Y. hunted for plants. Throughout the clear, cold night one white horse and three bays snorted at my white tent and raced over the chert gravel beds, but we saw no ghosts.

We wanted to reach a natural bridge in Barry County. People had told us we could not drive there, and we soon found eleven reasons why not: eleven big trees had been blown down across the road by the summer storm. There was nothing to do but leave the car parked where we encountered the first one. It was still a mile across the ridge and another mile down to the natural bridge. That was a short trip compared with the hike up Noblett Creek to the Spring Creek Canyon the day before. When X.Y. decided to go to a place marked on his map, he went in any manner possible. This trip was worthwhile.

The natural bridge spanned the hollow 40 feet wide. Recently the limestone roof had been falling down in big slabs. We found Walking Fern and a white-flowered *Sedum*.

After exploring the area around the bridge and the bottom, I turned back with a load of plants, moss, and rocks for our moss garden in Barrington, and started up the long steep hill to the car leaving trail signs to indicate I had gone on ahead. Soon X.Y. caught up with me. He was carrying a huge limestone slab which had fallen from the bridge. It was solidly bedecked with fossils. That slab, he figured, would be a base for Walking Ferns and *Sedum* which were growing on all parts of the bridge. This slab was to be added to our moss garden. He stopped to dig some Chinquapin Oaks to bring home as the seed burs were poor. Hornets chased him from the first one. He finally dug one and added it to my load since the limestone slab certainly was his load limit up the hill and another mile to the car.

That *Sedum*-covered slab stands on the north-facing slope of our moss garden as a constant reminder of that muscle-pulling armload from the natural bridge up the hill to the car. The white-flowered *Sedum* spread and spread.

At Roaring River State Park we could not relocate the *Tradescantia ozarkana* slopes found several years before. I had at one time found a double-flowered white one and hoped to find double forms in other colors. One big frustration in plant exploration is the changing and relocation of roads. You know right where the plant is, then when you approach the once familiar place everything is strange. The old road has been fenced off and abandoned. The only thing left to do is to take bearings to the approximate place and start walking. This we did, but finally gave up to wait until flowering time the next year.

From there we drove to Cassville for gas and groceries, then stopped along the highway to dry the blotters and eat lunch while enjoying the high bluffs beginning to color across the narrow valley. It seemed early for colors to be so evident, but the nights had been quite cool. After lunch we went on, passing many springs and creeks, then we began to find the country dry. We kept driving for miles only to find every creek dry.

The time came when we had to stop, water or no water. X.Y. finally found a spring across the road in a field, and I found a tiny spring in a small rocky branch near the car. McDonald County was proving to be quite dry. While cooking Spanish rice for supper, we were visited by a 14-year-old girl interested in plants. She asked the names of all the trees and herbs in sight. She reported that they had heard over their radio that by morning it would be 40 degrees colder than the morning before, which had been in the forties; and that the two and one-half acres of tomatoes they had planted for the canning factory had dried up, likewise the acre of corn and beans, and one and one-half acres of sweet corn. She was not merely curious about plants. It was a cow with a high-pitched bell that had brought her down to our car. She said she had never ridden a train or been anywhere, but would like to go somewhere sometime. Their six radios kept them up-to-date on the world.

We slept well and the predicted cold did not come; in fact the sun began to get hot early. At seven-thirty, X.Y. was on his way to look for plants while I broke camp. The girl came again looking for the lost cow. I was busy looking for lost keys. They had slipped out of my pocket on the road between the car and the branch. The dust was deep enough to bury the keys. I had just found the keys when X.Y. returned ready to leave for Roaring River Park.

At Arch Cave Hollow and all through the park we found wild Grape vines loaded with powder blue, sweet Grapes, large as domestic Concords. Treetops were bending with their load, and there was no film in the camera with which to catch the unusual abundance. Perhaps the vines had grown from the seeds of the cultivated plants since close to a sign, "No dumping" we saw piles of rusty cans ("*Juncus rusticanus*").

The day was spent in the southwest corner of McDonald County where X.Y. hunted near Pineville along Elk Creek which had solid walls of limestone bluffs. From there he hunted in backwaters near the town of Lanagan, and over a natural prairie on Highway 90. In that region he found again the wild Mock Orange (*Philadelphus pubescens*.) McDonald County is the only county in Missouri where it has been found. They can be grown from seeds, producing large shrubs with long drooping sprays of creamy white flowers somewhat smaller-flowered than the cultivated Mock Orange. It makes a very showy shrub in our Barrington woodland.

As we hiked over bare limestone bluffs we found Cloak-Fern (*Notholaena dealbata*). It was powder dry. After climbing over five hills we left. Already fall colors were claiming attention. Elm, Walnut, Hickory, and Sycamore in various shades of yellow; Sour Gum, Woodbine, and Sumac in their respective reds; and Dogwood showing red on the outer edges were the forerunners of the Ozark fall parade.

It was late in the year to see snakes, however I saw the first one of the trip near the top of a high bluff, a half-grown copperhead which made a strike from a distance, then ran. Very near by X.Y. found the cast-off skin of a large snake.

After shopping at Neosho we drove through Monett and Stockton, where we camped at Bear Creek. I collected moss-covered rocks and wildflower seeds. In a field along the highway I saw the largest *Gerardia* of my experience.

A field covered with Goldenrod and Blazing Star (*Liatris aspera*) presented a temptation too great to pass and we stopped by the roadside to eat lunch and change driers on the now many and swollen presses.

Later we took a long drive north to Merwin in Bates County near the Kansas line to look for the large Beard-tongue (*Penstemon Cobaea*) with large pale, lilac flowers. It is found only in this one county in Missouri, having crept over from Kansas. I collected a few seeds for our wild prairie in Barrington.

By now we were getting concerned about our house plants which we had left outside at home, though felt much better when we saw those of people in Harrison County still outside and showing no signs of frost. We went directly into Iowa where we followed the old Mormon Trail and through Davenport homeward.

We could turn our backs on Missouri for a few weeks and enjoy the comforts of home while anticipating one more trip to Missouri before the snow began to fly.

CHAPTER XII

SEPTEMBER

Flashbacks

On September 2 with the car packed to the rim, we set out for a Missouri collecting trip which was to continue for two weeks. We were hoping to add many X's to the county maps thus making the coming "Flora" more true to the area represented. Camp was made the first night on an isolated grassy glade, high over the Middle Fabius River where we were surrounded by Blazing Star and Goldenrod. To get water it was necessary to wind down a rocky trail to the river and carry the water back up the hill. Later we tasted water from a mineral spring close by and found the taste worse than the odor, however, hogs were drinking the water with no evident signs of disgruntle.

It takes many people to make a world, and while collecting plants along the highways and unfrequented byways in every county in the state over a period of more than 30 years we met many kinds of people. While breaking camp on Salt River in Ralls County, a man came to wash his car. The task which took three hours was lightened by a drink every 15 minutes. He yelled at me every time I went to spread driers in the sun, offering me a drink, which of course, I refused. At a distance X.Y. had been so intent on his work that he had not noticed the man, but he came forward when the man offered him a "jigger." He thought he was being offered chigger medicine and had to back out when he saw what it was. The man shied from all plants as poison, but took an interest in the insect repellent X.Y. showed him and suggested it might be good for a gnat that bit worse than a mosquito.

On September 7, we were given permission by the owner to camp by a spring on his farm. His eight-year-old grandson, a regular interrogation point, came to visit us. X.Y. said to the boy, "I'll bet I can tell you the name of that plant under the car."

The boy stooped down to look carefully under the car at the plant and answered, "I don't think you can. I don't even know it myself." Here was a typical case of where the possession of a little bit of knowledge, but not enough to know how little, deceives the possessor more than any one else.

At Blue Banks of Salt River we had a very beautiful view of bends and bluffs. While X.Y. got fossils from the bluff I got a sunburn from being in the water too long where the sun added extra energy by reflection from the water. The sun was not the only force in nature that could generate extra power on its own. When getting fuel for a camp fire for breakfast, I broke off a piece of limb from a dead tree lying near the camp. As the piece broke off, the attached section of the limb went, "zing," backfired, and punctured my face. Now I had a sunburned back and a punctured face.

In Pike County we got lost. A farm woman could give us no help, so we drove up and down north and south roads like the undecided fox that had been invited to two dinners and finally landed back at the starting place after the dinners were over.

The new moon peeped up over the hill to greet us just after dark, then dropped back over the same hill as slyly as he had come up. By now the presses were getting full. After a late supper X.Y. fixed my punctured face then worked on presses until 12:05 A.M. I often wondered how he could sit flat on the ground or kneel for hours in one position so intent on the job as to be oblivious to sensations.

In spite of his late hours, we got up with the sun the following morning. While eating our hotcake breakfast, the man who had directed us to the camp came by to tell us for the fourth time what a nice campground we had with wood and water so handy.

X.Y. went to examine the plant collection at the camp in the Cuiver River State Park. While I changed driers, an old lady came up the road whistling, to keep the "boogers" away she said. She was like the woman in Caracas who said she burned incense to keep the ghosts away. For once, I finished drying the blotters and replaced the warm ones between the specimen sheets and closed the presses before X.Y. returned.

We drove to Dark Hollow for the night, but everything looked too dry. The hollow was interesting, nevertheless, and we continued to drive at eight to ten miles per hour, easing over some very bad places at two miles per hour until time was running out and we had still found no water. We had to drive back to Hans Prairie and down to Auxvasse Creek for the night, where we camped on a nice level meadow by the creek.

The following day in Callaway County we drove to a farmhouse for water and directions. An old lady there told us how they made chill medicine. Take roots of wahoo, grind them, and mix with flour to make pills.

Fortified with our recipe for chills, we drove over a terrible road and down a steep hill to Stinson Creek. A man and two women came down the hill with rifles. They were hunting for squirrels and stopped to talk while we unloaded our equipment on a rocky point at the base of the hill where we expected to spend the night. For the rest of the day we hunted plants around the base of the hills.

At seven the next morning we started to hike over the hills, as we intended to explore Wildcat Canyon and found that it lived up to its name. The going was difficult over great boulders between steep sandstone walls.

The rocks were wet, and my leather-soled shoes constantly slipped on the wet rocks. Neither are leather shoes proper for hiking in dry woods. Dry leaves polish the soles like glass. We learned the best all-around footwear in summer for rough hiking is tennis shoes with good treads. They do not slip, are flexible, cool, can stand long hours of wading, and dry quickly with little damage to them. I scrambled along—rather slipped along—and saw one of the best canyons in the state. It is a canyon too beautiful to be hidden from the public. We returned in the afternoon to hunt plants on lower grounds. While at the campsite, the hill-family of three (now increased to 16) that had visited us the evening before came down the hill to fish. One young man cut poles for fishing. The mother said, "We would have serenaded you last night, but us poor hill folks don't associate with the rich lowland folks, and we couldn't find enough hill folks for a serenade."

They obviously took us for poor hill folks. This was a case of judging by appearances only. Should we consider it complimentary or otherwise? We settled for complimentary.

It brought to mind a humorous incident in our experiences where a man was judged solely by the clothes he was wearing at the time. A university-trained millionaire, author of many books, and member of top-ranking society was not permitted to enter a small, rather commonplace theatre because he was not wearing a tie.

A few weeks before, three totally illiterate Indians visiting the city for their first time, and wearing borrowed coats far too large for their below average stature of mature men, were permitted to go into the theatre. It mattered not that the coat sleeves hung to the finger tips—they were properly "tied."

Judging strangers by obvious appearances can be less accurate than judging plants by their environment only. The person being judged would necessarily fall into the concepts of the one judging. To the fisherman, anyone rambling along a stream is a fisherman; to the hunter, anyone rambling in the woods is a hunter, either of animals or medicinal herbs for sale.

The little boys who were along on this hot fall day were wearing felt hats and hot caps pulled down over the ears, for no apparent reason that we could see unless it was to keep off insects.

The salt springs northwest of Roach Port, a few miles north of Highway 40 had no vegetation in or along the water or flats bordering the stream. The water tasted of salt but had no taste or smell of sulphur. One solitary sandpiper at the edge of the salty branch was catching small forms of animal life in the water. He was so intent on his fishing that he let X.Y. come near for three photographs. Goldfinches also were drinking the salt water.

After supper we left our camp near the cave and went through a part of it. When we came out into the moonlight, X.Y. found a bat on his shoulder.

He had picked up a hitchhiker. At the camp where we were high enough to overlook the Missouri River, he set up the camera for a time exposure of moonlight on the river framed by bluffs. This trophy could not compare with my moonlight, island trophy.

We were spending a night high up a mountain. Near midnight when all was completely silent save for the low, rumbling, hair-raising concert of howler monkeys, I stepped out into the open which overlooked many miles of mountain terrain and faced a view probably never seen before, nor duplicated since. It is perpetuated only by a mental imprint on one's mind. A time exposure of that moonlight scene would have been unique. It would have revealed a fog, white as snow, hung at a billiard table level far below me. Rising like islands out of the white sea of fog, many mountaintops of various sizes, shapes, and heights, revealed details of forests and ravines in the full moonlight. I was tempted to wake others, and more reluctant to leave the scene knowing that never again would it be the same.

After X.Y.'s time exposure of the moonlit river was finished, we retired, though the pastoral night with horses, pigs, sheep, and a turkey with a bell on kept us awake most of the night. X.Y. found good botanizing along the railroad between the river and the hills the following morning.

We went to Fayette in Howard County for groceries and parked by the courthouse lawn. While X.Y. went shopping I did some mending while sitting in the car and listened in on the men already parked for the day on the courthouse lawn benches; the favorite assembly place of aged men.

One old man came along a little late and joined one who had been sitting alone on a bench not far from our car. They predicted the day would be hot—easy because it was already hot.

The first one said, "Ah's just been sittin' heah thinkin' to mahself ah'd like to cut some cohn. Haven't cut any cohn fo a long time. But ah's just been sittin heah a thinkin'. Ah believes ah'd like to cut some. It might be purty hahd on me. They say there's a lot of vines and weeds in the cohn. Purty hahd to cut. It might be hahd, but ah'd like to try it."

I was thinking I would like to find a cooler place to sit than in the parked car.

An old couple passed. He was constantly fumbling for the sidewalk with his cane. She stopped and said, "Honey, do you want me to put your glasses on you?"

He stopped shuffling and turned his face toward her, "Yes, Ma."

She carefully placed them on while he tottered on his cane. "There, now you can see the holes better," she said patting him gently on the back. They continued slowly and unsteadily on their way.

Later at the little town of Burton we stopped to ask directions on how to get out of town toward a place X.Y. wanted to reach. Five men holding down benches on the lawn each gave a different direction on how to get out of town. It was most confusing by the time each had given his own particular way to exit. We had to figure it all out for ourselves.

On return six hours later the men were still on the benches. As we paused for a light to change one called out, "Why don't you locate here? There's plenty of brush around."

X.Y. asked, "What do you mean?"

"Ain't you lookin' fer a location fer a C.C.C. camp?"

At the store the clerk thought X.Y. was looking for work in the coal mine at Blue Bluff. He was mistaken for anything that existed in the experience of the observer.

Blue Bluff where we spent the night was a beautiful spot, made even more pleasant by the owner who came down to bring us a big watermelon and two muskmelons. With the moonlight illuminating and softening the rugged beauty of the bluff, it was difficult to give up looking and go to sleep. Each half hour brought a change in the picture. As the moon changed position different ledges, shadows, and designs appeared. I lost count of the different pictures the moon artist painted on Blue Bluff that night.

By five-thirty the following morning we were up and were ready to leave camp at six-thirty. In Chariton County we drove along Missouri River bluffs to a river bend where X.Y. left me to dry blotters while he went collecting. A woman and a boy about eight-years-old in a little house across the road watched us park. First the boy came out and stood on the running board of the car asking questions. He asked if the car was the bread wagon. Then the woman came out to the car and, when she learned that we were studying plants began recommending some medicinal plants, of course thinking only that we were collecting for profit. She had a whole medicine chest at her door. There was Blue Ash for snakebite, but Tobacco was just as good; Plum leaves for asthma; and Wild Ginger for headache, though she had no remedy for what ailed the boy who was out of school because he was sick. The medicinal plants cited by residents scattered over the areas we visited would take care of all known ailments. Most of them were tested only by time and handed down by oral prescription from generation to generation and passed from neighbor to neighbor.

X.Y. returned before the middle of the afternoon, hungry after the very early breakfast and a long hike, so we drove back to Lisbon and stopped for his favorite chocolate malt to animate him for the next hike between then and lunch, which would be beyond Shannondale.

After lunch I again took care of things while he went on a plant search. Whenever he found a good place to collect, and rich in new records, nothing but an exhaustive survey or darkness would drive him in. He found such a place near Shannondale and came in late. After a good supper of mushroom soup, steak, mashed potatoes, green beans, fruit, and cookies, much relished after the long hikes, and eaten by moonlight and camp fire on the Chariton River banks, I read aloud to X.Y. from "Trails of the Ozarks" while he put his big collection of the day into press, and crossed off the new records on county maps. As he was about to run out of the prepared newspapers, I had to stop reading and tear a big stack of paper sheets for the day's specimens. On this night the moon mirrored itself in openings on the river. When the dishes were put away, I took some time out to walk down to the river bank to count and admire the many moons reflected from as many light pools up and down the river. Where a gentle breeze rippled a pool, the moon would ripple with it. I wondered how many other people had no more to do than to stand on a dark river bank and count moons in the river.

We had been out two weeks. Our vacation was over, and we must go home the next day. A vacation is always ended with mixed feelings. Constantly varying scenes day after day, highlighted by new plant records, could go on endlessly and are given up with regrets, however there comes a time after a few weeks, especially about dark, as we pass along looking for a suitable spot to camp for the night, when the lowing of cattle going to the barn; the calling of hogs by the farmer; the twinkling of lights in the windows; and other domestic preparations for the evening, spark a longing for the permanent dwelling, and we are glad to go back to our own home.

After a week of rest at home in Barrington, Illinois, we were ready to go back to Missouri. Leaving home at three-thirty in the morning on September 22 with frost on the ground we expected a cold weekend in Missouri, but found the day in northern Missouri to be hot.

The next day opposite Quincy, Illinois, on the Missouri side of the river, we parked the car near the river bridge and walked to search for a sandbar shown on the previous year's map.

We had a unique walk over tangled vines. For one-half hour we never touched ground. The tangle ended at the edge of the Mississippi River and no sandbar had been found. Later we learned that a dam had raised the water level and submerged the bar. Such is the life of a botanist who plans to write a flora.

While driving along the road with ditches on either side X.Y. saw an unfamiliar grass for this region. He drove along slowly, trying to identify it from the car. I said it looked like cultivated Rice. It proved to be just that when he stopped for a specimen. At Taylor we learned that Rice was being grown commercially along the river that far north.

The sandbar object of our weekend trip was negative and was crossed off of the map without results. We did learn something new about Rice.

There is a common tendency to think that a plant common to one region will not grow in some other place where it has not been found. History is full of examples where some plant was tried, out of its original habitat and found to be of great value in its new location. The Partridge Pea (*Cassia fasciculata*) is not found in our area in Illinois, but when some seeds were planted they quickly became a weed that had to be controlled. Since the flowers are attractive and since their nectar proved to be a favorite with honey and bumblebees; and the seeds, a favorite of pheasants in fall and winter, we even hesitated to thin the plants.

Weekend trips were the trying ones. They meant leaving home Friday after work in the city, traveling more than half the night and half of Saturday for collecting, and the last half of Sunday and Sunday night for return. If something special was found, the effort seemed worthwhile, but if nothing was found then the trip was crossed off as negative evidence and probably off the map for the future. On this trip we added Labor Day.

Early Sunday morning we set out in a light rain for a long walk down North Fork of White River near Tecumseh in Ozark County. Fishermen who had been on the move all night up and down the river seemed to be trying to fish the river clean of fish before the lake came in. They asked us as usual if "they were biting," meaning the fish, not ticks and chiggers. This had been a good Pawpaw year, and we ate as we walked. I think the white-meated Pawpaw has a nicer and milder flavor than the yellow-meated one.

Early Monday morning we started down Cedar Creek where we took pictures of Venus'-hair Ferns hanging under the blufs. This beautiful creek with its rare ferns and mossy bluffs was soon to be dammed. Along White River bottoms we again ate as many Pawpaws as possible. They also were to be drowned.

When the banks became impassable we followed along the slopes of the hill as best we could. Large and solid Oak trees, Hickory trees, and just any tree of considerable height had been cut and left to finally rot at the bottom of the lake. Thousands of board-feet of good lumber, a waste of natural products which man will some day have to account for, lay before us in mutilated acres. Catbriers (*Smilax Bonanox*) and others helped to slow us up as they caught in clothing or skin or blocked our path entirely.

The midafternoon temperature became hot. When we reached a low place where a ravine and spring branch entered the river, X.Y. went for a swim. The river was clear and deep here. Where I sat on the bank watching butterflies, dragonflies, and damselflies there was a large deep hole where the water was running swiftly toward the west on the side next to the bank, while out about 100 feet on the other side of the hole the water was running eastward. This was a large whirlpool into which X.Y. did not venture. Once pulled into that swift pool, even a rowboat or canoe would be in trouble.

Tiny gray blue butterflies lit on me getting moisture from my damp sleeves and shoes. One returned again and again to my damp hand. I could identify that one as he had a tiny chunk of moss on the upper tip of one wing. An old saying reminded me that if a butterfly lights on your hand you will get money. Others flocked in rosettes on any bit of moisture on the banks.

Water plants, at the margin of the shallow pools, were the playground of dragonflies and damselflies of many colors. The bodies of damselflies come in every color that nature produces. I watched this beautiful and busy scene until X.Y. finished his swim.

At three in the afternoon we started back to thc car and drove to a point on the ridge where on the map, it appeared possible to go down to Four Springs Hollow. When we reached the bottom of the hill, I saw only one small spring and much low swampy ground covered with interesting water plants and had turned back only five minutes before X.Y. found the large spring and tried to call me.

I took a short cut across an old farm, past an abandoned house, and was inspecting a big crop of small white gourds somewhat larger than golf balls. They were prospering by themselves long after all other cultivated plants had dwindled out in company with the crumbling house. There X.Y. caught up with me. The night was hot, brewing up a rain which drove us into the car at two o'clock in the morning. The next day we went home, having added little more than negative evidence.

CHAPTER XIII

SEPTEMBER

Poor Judgement Camp

A few years later on, September 23, we again started out for Missouri. The day was lovely, though at one o'clock on this Saturday we started out on the wrong foot. In the first place too much tension during the preparation between the Friday workday in the city and the present one o'clock had put me in no shape to travel—a migraine headache was no companion for a pleasant journey.

We arrived at Washington State Park at ten in the evening, where lightning and thunder were giving plenty of warning. Tumultuous thunder rumbled from the east and lightning building up to forked-tongue potential lit up the secluded, wooded park enhancing its loneliness at that late hour. Each following blast overhead which bumped and bounced away fading into the distance would immediately be followed by another which would build up to a hair-raising explosion. Instead of heeding the warning and stopping on high ground, solid with gravel in the unoccupied park area where clear water, shelter, and other facilities were available, X.Y. decided to drive down a road that dropped sharply into a field, and over a level road across the field to the river. *He wanted to camp by the river*. Between the field and the high ground of the park, a creek seemed to come from above the field where the river made a bend. I was worried in two respects. A heavy rain would certainly make the field impassable, and since the river was nearly full to its banks at present, much rain might send it out of the banks above the field where the branch could cut us off from the higher ground. But *X.Y. wanted to sleep by the river*. I made my bed on the front seat of the car, and X.Y. enjoyed the river bank for about an hour and had to come into the car. Right then we should have pulled to higher ground, but there were still a few minutes of that unlucky day left.

From midnight to seven-thirty in the morning it poured rain, and we knew we were stuck. I spent the night watching the river rise while lightning flashed out across it. That muddy field received its quota of nitrogen fresh from the aerial dynamo overhead while we were left to reap the reward of the thoughtless. We tried to pull out after seven-thirty when the rain had ceased, but scarcely went ten feet before the car mired down. I carried loads and loads of gravel one-fourth of a mile from above to place under the rear wheels. Three times, X.Y. jacked up the car to place gravel, sacks, and papers underneath. The mess we made in that muddy field was unforgettable and unforgivable. The river was rapidly rising.

X.Y. had eaten only a cold lunch and a cold supper, and I had eaten nothing the day before, and now there was no breakfast. Heavy with mud, hungry, and heavy of heart and limb, X.Y. finally gave up and hiked one mile up the hill to the home of the park superintendent.

The evening before he had said we would get up early and get to work as there was much ground to cover in the short time we had. Now at ten in the morning we had covered ten feet of ground, with much of the ground covering us. We surrendered to the mud, and when he returned from the home of the superintendent, X.Y. ate the same kind of thing for breakfast that he had for lunch and supper the day before, all the while wondering what manner of imp impelled him to leave the safety and convenience of the park in the face of a brewing storm to drive down and sleep one hour by the muddy Big River bank. That was no novelty—maybe habit had more to do with it.

At eleven o'clock a neatly dressed and freshly pressed young man arrived with a tractor. He smiled much, but said little, and pulled us out of the "Poor Judgment Camp" by eleven-thirty. We could easily see that even if we had been able to get out of the first hole we dug, there would have been one-fourth mile of such holes dug before we got off the field. Now the whole morning was lost. When caution goes, a downfall is inevitable.

About four o'clock after X.Y. had driven up and down the highway looking for a pond he never found, we arrived at Miner's swamp which we explored. Earlier in the year he had found Queen of the Prairie (*Filipendula rubra*) in Reynolds County, a relic standing alone in this county. It grows well and spreads in our Barrington garden where large feathery clusters of small pink, rose-colored flowers rise on stems three-to-four feet high, a very charming queen.

Not satisified at having mopped up the field mud, we went on to soak in knee-deep swamp water, looking for swamp plants almost until dark. X.Y. found a pink Cardinal Flower among a large swath of red one.

This time we made camp on an old rocky road near a small clear branch and there I cooked the first warm meal since Saturday breakfast in Barrington. Rain was coming again. We were safe for travel the next day, though had to sleep in the car. By piling presses, blotters, and stacks of newspapers in one wall the length of the car making a high partition between us we could have a long narrow cell in which to sleep on opposite sides of the "Collector's Wall." We were scarcely in our cells when rain began to pour.

On the road to Johnson County we stopped at Crocker to get groceries. About two mile north of Holden we hunted up one road and down the other searching for the white Closed Gentian. X.Y. was about to give up when signs became familiar. An abandoned side road used for dumping reminded us of our impressions of this place two years ago, before man defaced the beautiful natural slope. The farm to market road had been widened and blacktopped changing the appearance of the approach to the limestone glade, below which the Gentians had been found before flowering time. We now found many pure white ones and one tinged with blue. Since we lost much time finding the place, we reached Knob Noster State Park at dark.

We found no lights at the park, and I began to prepare supper by flashlight. The superintendent then came and fixed the lights, furnishing a life of luxury for that evening. We had shelter and no rain. It became hot and humid, a perfect stimulus for the myriads of crickets that chirped so fast I could not count the number per second.

Since we had so many modern conveniences at breakfast time, we allowed ourselves the luxury of a good and leisurely breakfast before heading northward to cross the Missouri River at Lexington. Near there we examined the newly-made sand mounds built up by the July flood, however found little of interest plant wise. We then drove eastward for miles through ruined land and wrecked homes headed for Wallace State Park for the night.

The temperature had suddenly dropped from 88 degrees to 40 degrees. The chill of the next morning must have affected our reasoning, as X.Y. put salt in his coffee and I put sour Milnot in mine. After breakfast X.Y. scouted for plants while I broke camp and packed. It was ten o'clock when he returned and we left the park.

After surveying wooded slopes west of Plattsburg we drove to Grindstone Creek where we ate lunch and changed driers. Leaving the presses exposed to the sun we went for a long hike over unburned, ungrazed wooded hills, ravines, and creeks. We called this the Indian Pipe trail and Coral Root haven because of their abundance. Indian Pipe or Ghost flower (*Monotropa uniflora*) is a saprophyte that pops up from a bed of Oak leaves and soon turns from white to pink and then to black. There also, we found many Yellow Lady-slippers. It was a rewarding hike.

North of Maysville we went to the top of the hill to camp above Lost Creek. We thought it would be warmer on the hill. Because of three Asters (one blue one, a possible cross, and one very prickly one), one very treacherous mud hole, and many deep, narrow cross ruts in the ungraded black dirt road, we missed getting supper over by dark.

Besides thinking it would be warmer on the hill, we thought it would be less frosty under a heavy growth of trees, close to some dense thickets. Frost had been predicted. In the middle of the night X.Y. woke me by calling out, "What's that thumping noise? There's something near the car."

We finally discovered it to be horses galloping in the pasture across the road. Before we could get back to sleep a big owl began to hoot over our heads, "Who, who, who, who are you?"

We explained who we were, and he moved farther away and began calling again. Another one answered with two more who's added. Each hoot owl continued to repeat the same number of who's over and over for an hour. X.Y. had gone to sleep when something began to tug at the covers near my feet. I called to X.Y. My call frightened the animal, which ran down the hill through the woods. It could have been a hog, a possum, a coon, or even a dog trying to steal a warm cover on a cold night. Most of the night was gone by the time we had peace.

During the night Jack Frost was on the prowl too. The whole car had turned white. Ice froze on the kettles of water we had ready for morning. Lost Creek proved to be a lively place to camp. We moved rapidly to get a camp fire going for a hot breakfast. The radio announced 43 degrees at eleven o'clock in the morning.

Not far from this "Galloping Horse–Hoot Owl Camp" we photographed the best display of Bittersweet we had ever seen—a Crab Apple tree completely covered by Bittersweet berries.

After changing the driers at lunchtime on the triangle between three roads, we set out across a hill along the river, went down eight hills, across eight ravines, and up eight steep hills where we found many Yellow Lady-slippers and Maidenhair Ferns. Eight such hills and hollows were enough for me, and I started back alone.

We prepared for another frosty night in an upland woods south of Albany. This place had even more owls, but the sky became cloudy and layer after layer of covers were peeled off before morning. The big camp fire I had planned for warmth during breakfast was reduced to the smallest one necessary for cooking the breakfast. The stack of hickory sticks were thrown back into the bushes.

Along Stillhouse Branch I collected Grapes on sandy creek banks while X.Y. botanized in the wooded hills. The wooded hills had been assured by the owner of a lot, full of squealing red hogs. X.Y. had stopped him astride the hog fence to inquire, leaving the poor man stranded on the wobbly fence with one leg outside and one inside the hog lot, while he talked.

Near Bogard in Caroll County we took a long walk through rich wooded hills to an artificial lake where an old lodge had once existed. To me the most interesting feature of this trip was the Kingnut Hickory. The nuts were the largest I had ever seen. I stopped under a big tree by the old lodge and cracked some, then filled the pockets of my jacket. One hour earlier the pale sun had lost the battle with Jack Frost over the black leather jacket which was by now a comfort rather than a burden. Added to his frosty accomplishments of orange-red Bittersweet berries and blue Grapes, were the fallen Hickory nuts enticing keen-eyed children for an after-school hour nut hunt. Half hidden and partly camouflaged by freshly fallen leaves, a nut sometimes revealed its location by a sharp tip or ridge of its shell. Any nut that failed to escape from the husk before falling, and lay hidden in the star-shaped husk, could start a race and scuffle among the children.

We moved on toward a place to camp while sounds of the nut hunt decreased directly as our distance increased. First we lost the sound of the rustle of leaves kicked about by the children, then the plunk of nuts thrown into the little wagon, then the sharp clank of nuts on nuts thrown into sacks, and then finally the gleeful yells faded into a shadow of voices.

We chose a campsite under a grove of Hickory nut trees where the nuts had already been gleaned. At duck we were settled when the children with their loads enroute to their homes passed and called out a good night. The revelry of the nut pickers had barely faded homeward when a drill began. Until far into the night, not far from us the drilling for oil monotonously thumped away deeper and deeper down into the earth. The night was comfortably warm and when the drill stopped, the stillness under the Hickory nut trees, broken only when an occasional tardy nut fell, seemed almost lonely.

Storms have a habit of coming up during the night, more often toward morning. On this morning a threatening one aroused us before daylight. It failed to materialize, but gave us an early morning start. We left the Hickory nut camp going southward crossing the Missouri River near Waverly where we took pictures of flood damage. Miles of farmland were still flooded leaving wrecked buildings, uprooted trees, homes with sides bulging, and watermarks above the doors. How long would it take to recover, and how long before another flood?

At Massa Creek, after lunch and blotter drying we went in different directions. I returned with a Water Plantain for which I knew X.Y. was hunting. I was secretly hoping he would not find it so I could surprise him. While skipping rocks across the glassy creek, X.Y. called from the car. He had already found the broad-leafed Water Plantain taking the wind out of my sails.

Farther down Massa Creek we parked at a sandstone area to examine the east-facing bluffs which were covered with moss, lichens, liverworts, ferns, and rosettes of Saxifrage turning the sides of the bluff to shades of yellow, grey, and green.

Near the top of the overhanging bluff with a drop straight down to the creek bed, a large blue racer snake lay in my path. I threw a dead stick at it with the sole intention of making it move out of my path. It almost turned somersaults as it headed down the slope and plunged over the bluff dropping through the air to the creek bed below. I could not see if the drop proved fatal.

We crossed a ravine and came up a steep bank out of the branch directly facing the dark green krinkled leaves of four Adam-and-Eve orchids and the widelobed Rattlesnake Fern in fruiting stage.

On returning, while going up a very long steep slope I got three breathing stops behind X.Y. and lost sight of him. He crossed to the bluffs north of the car. When I reached the creek I began wading to cool my feet as the day was near 90 degrees and then continued up the creek to a narrow boxed canyon where a waterfall blocked further progress. I decided to climb the bluff where X.Y. was calling about a grand view from up there. I got halfway up and found a perpendicular rock wall as high as a house and no quick way around, that I could see. There was a crack between two sections of the wall wide enough to go up as climbers ascend chimney-like precipices with one foot on one wall and the other on the opposite wall. There were a few bare vines to lend moral support though whether they were Virginia Creeper or Poison Ivy I did not determine. There were also a few projecting small rocks to cling to when they held. I got up perspiring as never before and vowed never again. By that time the view was still there, but X.Y. was somewhere unknown to me.

Later I found a way to slide down the talus slope to the creek and was skipping rocks and teasing a school of minnows when X.Y. finally came down the talus, rolling rocks ahead of him.

We drove farther down Massa Creek crossing the graveled bed of that stream 16 times down and 16 times back. The colored leaves of Maples, Dogwood, and Hickory were beginning to splotch the bottomlands, and red fingers stretched out where woodbines circled the limbs of large trees. Along fences Elderberry clumps over a foot across hung low with glossy ripe berries. Walnuts lying on the road were shelled by cars that rolled over them. Nature's food for man and beast was in abundance this fall.

From the beautiful Massa Creek area we drove to a creek south of Warrenton to spend the night. While I prepared supper X.Y. took a last hike and returned at dark jubilant at finding Goldie's Fern, new to the state; a Ground Pine; and a Leek on rich sandstone bluffs—a last grab of jewels.

A heavy blanket of fog added to the warm night. One of our ever-present friends, the big hoot owl with eight syllables of one pitch, called much of the night to another one more distant. The little screech owl also quavered over our heads. We gave the big hoot owl a gold star for good attendance. He seldom missed a night. No camp would be complete without his broadcasting. He always knew the questions, but never the answers.

We left for home before sunup the next morning, and for the first time in our Missouri exploring history crossed the Mississippi River at 8:15 in the morning, standard time, and arrived home in daylight.

CHAPTER XIV

SEPTEMBER

Rescue of Doomed Plants and a View of a Juniper Grove

On a clear frosty morning on September 24, we left Barrington at seven o'clock to make a fall survey for any plant that might be added to the "Flora." Our first destination was Cook's shelter house in Reynolds County where we arrived in late afternoon. There along the spring branch we found Eastern Witch Hazel (*Hamamelis virginiana*) beginning to bloom. This one is restricted to five adjacent counties while the Ozark spring-blooming species is well spread in the Ozarks.

Fall was painting the Sumac red and maroon-red, and the Hickories yellow, then cleaning the paint pots and brushes off by splotching and streaking yellows, reds and mixtures on Sassafras and Sugar Maples.

Leaving this charming spot the following morning, we climbed the hill and drove to Bunker to a good swampy meadow. Grass-of-Parnassus and many Ladies' Tresses orchids (*Spiranthes cernua*) spread over the wet meadows. This fragrant orchid has a spike four to five inches long, with flowers opening first at the bottom of the spike. The bee that fertilizes the orchid starts with the older flowers at the bottom and spirals up with the flowers around the spike, like the little brown creeper bird, following the flowers upward to the top, picking up pollen on the way up. Then again, like the brown creeper, the bee flies to the lower flowers of the next orchid plant spreading pollen as he spirals.

The rare, Marsh Bellflower (*Campanula aparinoides*) found only in this area in Shannon County is another relic left by the retreating glacier. The bells on this plant point upward while the blue bells of *C. rotundifolia* more or less hang downward.

A man at the little post office at Hardage gave us a remedy for poison ivy. He said, "If you are sensitive to Poison Ivy, you are also sensitive to about 70 other plant poisons. The best remedy is to go to a library and get all the books on the subject, and stay there and read them, thus avoiding the plants."

That was good advice, however we already had the poison and were looking for something to relieve it. X.Y. was also hunting for a new highway from Bunker to Highway 19. He thought we had started out on it. It actually was the beginning, though it ended too soon. We spent two and one-half hours going seven miles, crossed the graveled-bottom creek eight times with fingers crossed hoping the gravel would hold us up. Four times we crossed spring branches, following the marks on trees indicating the road was to be there and hoping each time we went around a bend that we would find a stretch of new road.

From Pontiac we drove down a washed-out road, eventually to be under water from the dammed White River. We arrived at a place to camp after dark, and I cooked supper practically without seeing it. Though many such suppers were cooked "blind" no complaints were added because of lack of proper preparation.

The following morning X.Y. went up the river while I broke camp. I thought I could overtake him and kept "yoohooing" trying to locate him unaware that some cataracts and low waterfalls near him drowned out my voice. When I reached the thundering river sounds, he was far ahead. I stopped to admire the falls and rest. Right before my eyes I spotted a new plant that brought from me ejaculations loud enough to put X.Y.'s war whoops at new discoveries to shame. In a swampy spot near the edge of the water between a Cardinal Flower (*Lobelia Cardinalis*) and a Blue Lobelia (*L. siphilitica*) stood the most beautiful hybrid of the two. The spike of flowers was a shade of clear rose-purple resembling the more delicate flowers and leaf structure of the red Cardinal Flower, taking only the blue color from the blue parent. This plant was too beautiful and too rare to be "dammed" so I dug it up with as much soil as possible, wrapped my handkerchief and large plantain leaves around the soil, and hastened back down the river to the car hoping to keep this rarity alive until it could be planted in a safe place and perhaps bear seeds. This part of the river was soon to be under 90 feet of water, the falls hushed, and rare plants smothered forever under muck.

This was Lobelia day. At Bratton Springs which bubble up out of flat ground amidst Water Cress, we followed down the spring branch finding Blue Lobelia, a lavender form of the Blue Lobelia, and finally we came upon a pure white form of the blue. Also, this spring and the white Lobelias would vanish.

At the next stop on Little North Fork of White River we walked down the river to take pictures of the pendent Maidenhair Fern or Venus'-hair Fern (*Adiantum Capillus-veneris*). It was necessary to stand in knee-deep water to photograph this delicate fern hanging under the dripping, mossy limestone bluffs. While it is found in such habitats only in the Ozark sections, the common palm-shaped Maidenhair Fern (*A. pedatum*) is found in almost every county of the state except the four most southeast counties.

That night we had one of the nicest campsites on flat rocks at the very edge of the clear running stream where the moon lit up the gurgling, winding river, and the fireflies lit up the silent, misty fields. The circle of bushes surrounding our camp fire danced to the flicker of the camp fire flames. At such campsites free of mosquitoes, rain, and the noise of civilization, for a short time there is a longing to be primitive again. We let the river current carry away past, present, and future worries while we relaxed in the glow of serenity, man's ultimate aim of existence.

We wanted to get a last look and make a last survey of Big Creek. The car was left on top of the hill, and we walked down a very old road to the beautiful creek, much too beautiful to be buried. We took some pictures of Big Creek and of a huge rock in the creek which very much resembled a boat stranded in the water.

After awhile I walked back toward the car pulling myself up over steep places by a wire along the side, put there for a boundary line and not as an aid for climbing. At one place in the woods which would be below the water level when the lake eventually covered the area, I saw the entrance to a cave. So much of the Ozark region is hollow with caves. This raised the question. Would not the lake water enter the cave openings in many places and fill all connected subterranean cavities, and eventually come out at places where springs emerge?

In Monett, Missouri, where a deep well for the town water supply had been bored into a cave under the town, a bushel of apples poured into the well came out at Roaring River Spring 25 miles away.

Before reaching the car on top of the hill I entered a grove of Red Cedars (*Juniperus virginiana*) covering acres. Here I lost all sense of time and direction. I was soon in a strange, new world of nothing but cedars. Each cedar was a perfect tree with compact branches hugging the ground, gradually tapering in conical form to the topmost tip. Not another thing grew there. The outer branch tips were spaced about four to six feet apart. A thick carpet of needles through which nothing else could grow had accumulated over many years. I wandered on and on through the cathedral of green with their pointed tips now touched by the setting sun. The evening wind blowing softly through thick branches of so many cedars, played a variety of melodies which were not disturbed by my steps on the soft carpet of needles. Except for these lovely trees only the sky above was in sight. The awe experienced in this cathedral of cedars was partly due to the absence of any other single object to distract from the majesty of the perfectly formed trees; and partly due to the particular lighting and shading at sunset; and the variations of soft music blowing through the trees. Surely this flat area at the top of the hill was a perfect place for them. Like X.Y., I would have given much to save this stand of trees, at least 275 years old according to known ages of trees of that size. The dam would bring the lake up to where the bluff and hilltop would become real-estate projects ending the lovely cedar grove (later it did become a preserve).

This was a unique experience. I wanted to walk on and on around one turn after another. But night was coming on; X.Y. would be returning, and there would be no supper ready.

I did return ahead of him and was confronted with the problem of heating some food. Our Coleman stove was not working. I was back to earth with problems to solve. There were no large rocks in sight, besides I did not want to take a chance of the wind playing tricks with a fire on this exposed grassy area. Near the car a trench one and one-half feet deep and less than two feet wide had been dug. The grate we always carried just fit over the trench, and a little fire was made in the trench under the grate—just enough to warm some canned foods.

The following morning after breakfast we drove to within one and one-half miles of the Arkansas line in Taney County. We left the car at Brown Ferry and hiked over the bluffs. X.Y. was looking for a plant which he had sought for years. I hiked alone up an old trail, photographed the river, and found *Stenosiphon linifolius*. The flowering spike four to five feet high was covered with small, lacy white flowers. Neither of us had ever seen it before. It has been found only in two counties in Missouri, Ozark and Taney. X.Y. found the plant for which he had hunted so long—French Mulberry (*Callicarpa americana*). This was the first time it had been found in Missouri, and the last time, since its location was soon to be covered by the lake caused by Table Rock Dam. The fruits of the French Mulberry are beautifully colored from violet to blue-purple. It makes a nice cultivated shrub.

Our camp that night was high and dry on top of a bluff. On such a cold starry night, it was pleasant enough to cover the head just in case the big hoot owl, hopping lower and lower in the branches of the tree overhead, might mistake a movement for the movement of a mouse or rabbit, and come pouncing down with his powerful claws.

Near sunrise a dense fog hid everything. We had no water to wash the dishes, but the fog had almost washed our beds. It was necessary to dry clean the dishes and dry the beds.

We were on the hunt of a Pink Dogwood (*Cornus florida* f. *rubra*) near Cedar Creek. We went back and forth three times over the same road before we located it. During the spring we had found it in flower and now wanted to get some fruits to grow, hoping that a plant grown from seeds would adapt more easily to a different environment. Also we wished to see if some or all of the seeds would produce pink flowers. Since there were several trees in that vicinity producing pink flowers, we suspected that pink was inherent in the seeds and not a product of one freak tree.

In Ozark and Taney Counties where so much of the bottom farmland was transformed to lakes by the dams on White River we heard many laments from farmers. At one place a man said, "The senators told us,

'Only you farmers are to blame. You thought you'd get rich when you let the dams come in.' One farmer started to raise a crop of oats, and they made him get out: he hasn't been paid for the farm yet.''

We went again to bluffs where we had been in the spring along White River, went down a steep one and back up along the north-facing bases where X.Y. found Celandine Poppy, Squirrel Corn, and many other nice plants which were soon to be flooded out by the dam.

Ice on the windshield of the car the next morning suggested that it was just as well that our vacation was over. Thoughts of home almost 600 miles away began to speed up our departure. It had been a good trip with many new records, and some rich memories stored up for enriching later life. Some trophies had been collected for my mental garden: the lobelia hybrid which never can be seen again; the pendent maiden-hair picture clear in my memory long after the mud has silted over it; and the juniper cathedral which years later could strike awe. All are deposited, ready to be pulled out from their mental casements for review.

CHAPTER XV

SEPTEMBER

Rainbow in the Morning and E.S.P.

One year later on September 17, we drove into Missouri through Hannibal. We left the alarm clock at the Little Grassy Creek where we had spent our first restful night. We had no intentions of cluttering up the landscape with an alarm clock. X.Y. had put the clock in the grass by his head. Whoever found the clock later could see no other evidence of occupation than a depression in the grass obviously not made by a horse but with an alarm clock. There was no need for an alarm clock the next morning. The night had started out bright with northern lights streaking across the sky, which we elected to watch. Soon after midnight lightning flashed constantly across the sky. The most spectacular, though not the most conducive to sleep, was the metal bridge, the whole framework of which lit up one side of the river to the other with each flash of lightning—a framework of fire. Nature's dynamo sparked the metal bridge with more kilowatts than might be possible with the coming hydroelectric power plant on the "dammed" White River. The night was hot, and neither of us slept.

Though I did not attempt to count the seconds between the flash and the thunderclap, it was evident that they were close together. Farmers nearby may not have been entertained by the luminous display of the bridge; nor conscious that nature was spreading nitrogen by means of the raindrops to fertilize their crops.

The sky continued to exhibit a perfect rainbow in the west early in the morning which called to mind, "Rainbow in the morning, sailors take warning."

It did begin to rain at our first stop. X.Y. started out in light rain which soon developed into a downpour lasting two hours. He returned soaked through and left again after leaving his collection of plants at the car.

The smell of rain has local color, depending on where it rains. Not of rain itself, but of smells carried by raindrops and wind. The nuclei of raindrops may be pollen, spores, smoke, or dust. Large wind-blown drops now loaded with suspended clay particles picked up from the air above the newly-graded road wafted a damp smell of dust in through the crack I had left above the windless side of the car. When the air was cleaned of dust, cool, refreshing smells of leaf mold came up from the forest at my right. A shift of wind from the field behind me brought the pungent smells of September—dried hay spiced with Goldenrod.

After the rain ceased I tried to walk up the road, but the clay came up over my shoes. It was better walking in the woods for one hour until X.Y. came. The silence of the washed woods was broken only by moisture dripping from the leaves. Even the birds had not ventured to chirp a note. There was a Slippery Elm hidden from sight somewhere in the forest asserting its presence only by a slippery tang in the air. A feeling of comradeship with the woods always transcended any fear I might have through walking the forest alone.

When X.Y. returned, a man who came by in a jeep told us we could not possibly make it up that road. We took his advice and turned back to another road. The road we took was no better. X.Y. could not make the first low hill and backed down toward the junction where he slid crossways on the road. We piled branches and brush in front and back, and continued to sink in the mud. While getting brush, I disturbed a hornet that sent me out of his premises with a sting on top of the head. On the retreat one shoe pulled off in the clay and that foot came up with a clay cast. We were in a mess. Finally an elderly squirrel hunter came along. He must have had experiences in such dilemmas, because he threw away the brush and manipulated movements that finally got us out though the exhaust tube had filled with clay.

We had no sooner got out of that, when a man came by in a car saying he had come over the road we had first wanted to take, and he could take us back over it. We followed him for six slippery miles of misery. At the first steep hill he stalled and backed down for a second start, made it and waited at the top. I got out of our car and X.Y. made a run for the hill. My presence in the car at critical times made him nervous. He needed to give full attention to driving. He did some terrible side slipping within six inches of a deep ditch on one side, where for one terrible moment, I feared the car would leave the road. More rain was threatening. I slithered up the yellow clay road as fast as possible. What a time to be traveling on newly graded unfinished roads. By now I was beginning to believe in the morning rainbow. Early that morning we had seen a bright one. The belief in rhythmic patterns of earthquakes or hurricanes is based only on coincidence, however, "Rainbow in the morning, sailors take warning," is founded on scientific facts rather than fiction.

Back to the highway we headed for Mercer County. Terrible black clouds rolled, lightning zigzagged and sent thunder rolling from cloud to cloud.

On a highway where work was being done, a workman motioned for X.Y. to come on that the pass was clear. Just as he motioned X.Y. suddenly started backing down the hill. The expression on the boy's face was something to see. X.Y. had seen a clump of white Thistle and as usual backed down to get it. We often puzzled the inhabitants of an area by doing things contrary to their line of thinking. The boy, of course, had not seen the Thistle, and if he had, to him, it was only an obnoxious weed.

Near Saline we drove over a newly graded road to Conner Branch to camp. The road was under construction beyond that, and not at all passable. We pulled off the road about 50 feet from the creek, which was only a tinkling stream at the bottom of its steep six foot deep banks. We cleaned up, removed the clay we picked up enroute, and rolled the clay-covered shoes in newspaper. X.Y. put on all clean clothes and went to talk to a farmer nearby while I got supper. Another storm was threatening, therefore, after my patience at waiting for him had worn thin, I ate and washed my dishes. X.Y. continued to talk until a downpour began and soaked him through again, then he had to eat in the car. There was no time to bring our kitchen equipment into the car.

From dark until ten it poured. We were prepared to sleep in the car feeling safe from the terrible lightning. A car is the safest place to be in the case of lightning, though the worst possible place to be in time of a tornado which can pick up a car and toss it around like a matchbox. Since there were no large trees over us we were safe from the wind. I feared the worst for our stove and icebox and other things left outside the car. The creek never entered our minds. Having had no sleep the night before, we slept well, however, if we had seen the creek rise eight feet and over the bridge forming a swift river to within a few feet of the car, we might have lost a few hours. Our equipment left outside was covered with sand and mud. The man at the house said his rain gauge showed four inches had fallen during the night. The rainbow in the morning had given fair warning.

At Crowder State Park the next night X.Y. had the luxury of electric lights and a picnic table on which to work. After his work was over we toasted marshmallows and pretended we were having a picnic as others were doing at another building. They left at ten-thirty, and we remained.

Again it poured rain as we spent the night one-half day's drive from Jamestown, south of Columbia. Our car had not been in the Columbia area for this trip.

Near Sandy Hook within a few miles of the Missouri River we stopped on top of a hill to eat lunch and dry blotters. While sitting there I felt that we were being watched. I said the same to X.Y. He replied that people who passed were always curious and watching us, which was true, but I felt that this was more than curiosity. E.S.P. often takes over against my conscious reasoning.

An elderly man came across the field, and X.Y. asked him where we could find some sandstone outcrops along the Missouri River. X.Y. had indications from the map that some existed in the region.

The man said he had some on his place, but it had a government mark on it. X.Y. wanted to go with him to see what plants were there and the old man said, "I'll take you there, but you leave your knife here." X.Y. unstrapped his knife and left it. The hunting knife so blunt from digging plants that it was no more a menace than a stick or a rock. It was surprising that the old man trusted him even without a knife.

As they started to leave the old man turned and remarked to me, "You'll be safe here on the road." That was a strange remark to make to me. Never once had I ever thought of not being safe alone any place, even in the woods in all our wanderings. No one had ever suggested otherwise, and no one I had ever met ever said more than "Howdy" in passing.

I struggled along trying to dry the blotters on a hill in 49 miles per hour wind, chasing them all over the muddy field for blocks, and trying to weigh them down with rocks and sticks. People kept passing and staring. Again E.S.P. popped up.

When X.Y. returned alone we drove to a dead-end road by a house where he knocked on the door to get permission to walk across the field to the supposed sandstone bluffs above the river. As no one answered the door or his call of, "Anyone home?" we parked the car and started down across the field toward some woods.

Soon we saw a man come from the barn and start following at a distance of about 500 feet. I remarked again, "X.Y., there is something fishy going on in this neighborhood. We are either being watched or else people are guarding something."

"You're imagining things," was his reply. Right then he had only a sandstone outcrop on his conscious mind, and my subconscious was working.

At that time I was making some study of plants that grew, or those that were never found, under Walnut trees. I carried a notebook and some equipment for collecting the specimens.

We found limestone bluffs and some big Walnut trees, but no sandstone, and after a little more scouting around, we returned. As we went up the last hill toward the house we saw two men looking under the hood of our car. Three other cars were parked there. An armed state trooper, and the armed county sheriff were looking through the car. Four farmers with shotguns on their shoulders stood near. My hair stood on end.

As we approached, the state trooper introduced himself. X.Y. said, "My name is Ramble, X.Y. Ramble. Here, I'll show you my name," and he reached for his card to show the trooper.

"Yes," the trooper answered without looking at the card. "It's spelled R-A-M-B-L-E," then he added, "You live at 117 North Hill Drive in Barrington, Illinois."

X.Y.'s eyes popped, and all he could say was, "What?"

The trooper then said, "I called Springfield and found out through the car license department. I was just about to call the police in Barrington to ask about you."

X.Y. said, "Well, I don't understand, I only went down to look for some sandstone bluffs I thought were there. I'm studying the plants in Missouri."

"I could have told you there are no sandstone bluffs in this neighborhood," the trooper answered. "I studied geology at Washington University."

"I had geology there, too," X.Y. added.

Then began a session of times, dates, teachers, and classmates in common, while the sheriff and farmers shifted from one foot to the other and relaxed hold on their guns. He was soon convinced that we were innocent of the robbery that had occurred the night before. An old couple in the neighborhood had been robbed of their life savings by someone in a car with a Chicago license.

The farmers went on their way, the sheriff on his way without a prisoner, and the trooper, now a classmate friend, escorted us back to the highway. Now in addition to believing in morning rainbows, I am beginning to believe in extrasensory perception.

What a joke on us it would have been, if we had looked longer for the sandstone, while the trooper called the Barrington police.

One special aim of this trip was a visit to Elk Prairie, about ten miles south of Rolla, one of the few virgin prairies left in the state. Mr. F. let us enter and park about a mile from the prairie. For three hours we walked over the prairie seeing thousands of Ladies'-Tresses orchids, 38 clumps of Downy Gentians, and many records new to the county. X.Y. considered this to be the finest natural prairie in the Ozarks and one of the best in Missouri. If he could finance it, he would buy the prairie, a typical upland sinkhole pond, a Bogbean bog, a Holly marsh, an unusual canyon or two, and a few other rare places in the state for preservation only, and I would buy the Cedar grove.

They represent such a small percentage of the land in the state, yet such a high percentage of aesthetic and historical value. It seems a crime to destroy what can never be replaced.

One more place remained to be visited before starting home. About four miles west of St. James was the old homestead of my grandfather which had existed by the creek that bears his name, Cox Creek. X.Y. remembered plants seen there and wanted to see if they still remained.

We found the creek, and a swampy spot where the spring near the house had been. Remains of the old Matrimony Vine were still around the spring site, though not even the stones of the foundation of the house could be found, or the stones behind the house where I had sat as a child with Hollyhocks as comforting companions after being punished for wading Cox Creek in my new patent leather shoes then skuffing them through the road dust. My three-year-old philosophy was why make such a fuss about shoes, why not buy a new pair. The Hollyhock flower dolls with pink or purple flaring skirts understood. The lawn and garden was now a field. Not a sprout remained of the big White Poplar with gray spreading branches which held the initials of every child young or old in the neighborhood. These big branches had borne the weight of everyone not too dignified to climb. The huge Walnut trees, and Hedge Apple, or Osage Orange (*Maclura pomifera*), trees along the road to the creek were gone. The Osage Orange which had been used as a fence was now replaced by wire with clean fence rows. Some one in more recent times had felt the same about fence rows as the old Frenchman who lived across the road from my grandfather. As a child, I thought the two men were quarreling when they argued about fence rows. The Frenchman protested loudly that space for growing crops on my grandfather's place was being wasted. His were clean. My grandfather argued that fence rows belonged to the birds and mammals. The banks of the creek now were cleaned of their lush vegetation formerly protected by grandfather as a natural homesite for the birds.

At any future visit to a stream where Willows grew, the sensation of their fragrance, though less keen than as a child, always transported me back to our childhood rambles up and down Cox Creek. Had our grandfather suspected that we peeked into every bird's nest, our rambles might have been curbed. We kept lookouts.

At ten-thirty on that bright sunny morning, after eight days out, we headed for our home in Barrington.

CHAPTER XVI

OCTOBER

Frost Flowers, Spice Bush Surprise, and Bears

We hoped to make a few fall trips before winter brought a halt to all chances of adding any more records to the "Flora."

Incidentally, on this particular frosty night of October 17, we picked the banks of Warm Fork of Spring River for our campsite. We happened to arrive there at dusk from our home in Barrington, otherwise it might have been a spot with a more timely name. To save time we seldom stopped until darkness prevented further work. If, in some of my moonlight sallies for firewood, I had picked up a snake for a stick, the fire would have started right there. Cooking by flashlight became a task. If, only, at such times, one had three hands. Holding a flashlight under one arm or between the knees or slung from a belt while stooping to prepare food over a camp fire is tricky to say the least. Looking into a kettle against the light of a camp fire to see what is going on within, yields about the same results as looking into a deep well—just a black cavity. The flashlight must be at hand again for a peep and a stir—that is, if you have not put the flashlight down somewhere and cannot find it. In the meantime, the pot boils over.

But not so on this evening. The small, smoky fire of soft bottomland wood lent barely enough heat to limber my fingers, heat a can of pork and beans, and one of spinach. From Dittany, a mint collected earlier on the road, I made a cup of hot tea. We ate quickly, and without troubling to scout around for more wood, I made short work of dish washing in warm water. By that time the moon furnished the light.

We crawled into cold beds at seven-thirty. My cold feet were forgotten for a moment when the usual, "Who, who, who, who are you?" hoot owl serenade began east of us. From the low hills at a greater distance west came the answer of higher pitch and one more who, "Who, who, who, who, who are you?" This continued by the hour as the moon rose higher. Slowly the "Whos" approached each other. It was too early to sleep and I was cold, yet I had entertainment with the owls for a radio, quite varied and hair-raising at times. By making a small hole through the covers, television came into view. Perched on large limb of an old Walnut tree was Capella. She began to climb. I called to X.Y. who had gone to the car for an extra blanket to bring me his watch to put under my pillow. I wanted to see how long it would take her to climb to the top of the tree. To what form of accomplishment does one aspire deliberately to keep awake until a star has finished climbing a tree. At last she reached the tip-top of the tree and I could go to sleep—that is if I could get my feet warm.

There was some restless sleep between the time the hoot owls gave up and the black-crowned night heron began. Before daylight the heron began his weird croaking or better still, quawking, since the common name for the bird is "quawk" from the call he makes. He seemed to stay in one place and I took careful note of the direction from which the quawking came, hoping he would stay there until I was ready to look for him.

At daylight I got up and began a double search for firewood and the heron. Following "quawks," now, further apart as day advanced, I crept through dense growths of underbrush and looked up to see the heron's polished black feathers and red eye reflecting the early light. There I stood motionless to watch the movements accompanying the "quawk," then proceeded to get the wood with which to cook coffee, eggs-a-la-ranchero, toast, and to dry my soaked feet.

Before reaching camp I had been sidetracked by frost flowers. The night had been ideal for the growth of frost flowers on *Verbesina virginica*. Few people get out early enough on a frosty morning to see these elusive fairies. Like Cinderella they evaporate before sunrise. Given a fairly warm moist soil, a sudden frosty night, a stem of *Verbesina*, Dittany, or other member of the mint family with dead outer stem surface, and active root growth, the result may be pearly-white frost flowers in icy design. Just above the soil level around the plant stem, Jack Frost works in three dimensions. Root pressure pushes moisture up the stem. Just above the soil the stem bursts—the more irregular the cracks, the more complex the flower. As the sap oozes out of the cracks, it solidifies into thin fluted ribbons an inch or more wide which twist and curl to form a flower of tissue-thin ice and air as elusive as cotton candy. They were often the cause of fall tardiness at school. All along our road to the country school grew Dittany which produces beautiful frost flowers. It was much fun to pop them into the mouth and instantly find nothing but a cold tongue.

A pang of hunger brought me back to the realization that something more substantial was necessary than a cold spot on the tongue. After sampling a few frost flowers, I picked up the load of wood and hastened back to camp where a noisy billy goat was busily inspecting everything. X.Y. with one-half of his face shaved stood close to the car door hurling sticks at the intruder.

At last a fire glowed, "billy" had gone homeward, and breakfast was ready. Toast from a pop-up toaster may be convenient, but nothing compares in flavor with toast made over live coals from oak or hickory wood. We sat on newspaper with the fire at our feet, and the rising sun at our backs, and breathed the clean, frosty air. Not one thought of responsibility or even mild nostalgia for the comforts of home troubled our minds.

As usual, now that frost had accomplished some good and some bad, we could expect Indian Summer. Does nature carry out a purpose when frost kills the garden, and fragile flowers, then runs a long period of golden sunny days, softened by haze-lazy days free from plaguing insects? That is a frequent fall pattern.

After breakfast X.Y. took a short hike to the hills nearby while I broke camp and heated a kettle of water for a personal touchup. By nine o'clock we were ready to leave the "Hoot owl, Night Heron, Billy Goat, Frost Camp" for another area.

Fall colors in Oregon County were not very good because of too much drought. Along Frederick Creek, while X.Y. studied plants, I scouted slopes near the top and picked up moss-covered rocks, lichen-covered rocks, and pitted rocks of various shapes and colors to be used in a terrarium display at home and in our moss garden, the only limiting factor in such collections being my own carrying load limit and the car capacity. I select lichen-covered rocks with much respect. Usually it became a game of select and reject, repeated with a few musts and many discards before the slopes bristling with more rocks than soil was explored, and I stumbled downhill with my load.

We next went to Alton, for groceries and found the black-eyed peas I had been asking for at all the country stores in the south. Here, I found them freshly hulled and tender. From there we drove to Hurricane Creek, now bone dry. X.Y. hiked while I ate Persimmons and Grapes, and gathered more rocks including "blossom rock, or sugar rock" as it is sometimes called.

Our native Persimmon, a North American member of the ebony family probably of more ancient origin than the blossom rock nearby, survives now in the southern half of the United States while according to fossil records, it spread widely north to Alaska and Greenland far back in Cretaceous times. Drastic changes in climate since that faraway time have squeezed it into a more restricted area suitable to its needs.

Quartz crystals of the blossom rock glistened in the sun like diamonds all over the slope. Slightly tinted quartz crystals formed rock patterns varying from the fine-grained, white crystals bunched like florets and combined into a cauliflower-like head of rock to large dark-rose or purple crystals forming reflecting facets on concave or convex shapes of rock.

Enterprising people have built up a business by transporting the rocks to roadside markets. Such rocks exhibit themselves best in their natural environment. At certain known areas an entire hill slope of such rocks against the rising or lowering sun can stop the most dull-witted, dull-sensed, unconcerned hiker in his tracks.

This was a day of many short hikes ending at the West Fork of Fourche Creek to camp at "Dry Creek, Hoot Owl, Meadow Lark Camp." Wood of a fallen Honey Locust tree supplied a good fire for cooking the black-eyed peas so long anticipated. Remembering the previous frosty night, I selected a large, clean flint rock from the dry creek bed, heated it in the camp fire coals, and wrapped it in many layers of newspapers. This would keep the bed comfortable under frost-covered blankets until morning.

Thanks to the hot rock I arose at daybreak the next morning brimming with energy. I was starting a fire when a voice from a frosted mound on the ground came out of an eiderdown bag, "How about hotcakes? Got any honey?"

Hotcakes and honey was a sound I liked to hear on a cold morning. "How about scouting up and down this creek for some dishwater?" I replied. "Do something to earn your cakes."

One common factor in our constantly changing campsites was what one had, another lacked—seldom 100 percent perfect. After breakfast we sat in full sun that paled the flame of our camp fire, and ate in easy reach of the coffee pot and stack of hotcakes kept warm on a hot rock by the fire. A meadow lark sang out across the frost-covered field. Before our breakfast was over, the warm October sun had changed the white field to one of green with white frost shadows of trees and shrubs growing shorter as the sun rose.

Going eastward, we stopped at Doniphan to buy steak and apples to go with mushrooms found earlier in the woods. Near Naylor we camped at "Birthday, Steak and Mushroom, Cotton Gin Camp," and slept late the next morning. After a late breakfast of oatmeal we began an eventful day, four miles south of Naylor. We drove over bumpy dirt roads eastward to a spot to be surveyed. Bumpy dirt roads always proved to be the worst roads for writing up notes in motion. The short staccatos on notes produced by speed on good roads became crescendos on bumpy roads at low speed. They, however, furnished an on-the-spot description even though they had to be rewritten later.

We left the car at the dead-end of a dirt road and cut across a weedy field to a wooded area that had been a swamp earlier. Leaving the swamp we crossed a sandy field where peas had been harvested, then crawled through a fence into another sandy field with large watermelons of 20 to 30 pounds each. Blackened vines reminded us of the frosty night before, and suggested that the melons were not what they seemed. Beautiful melons, dark green and light green stood out boldly by the hundreds now unscreened, since their foliage hung black and limp close to the ground. We stared in amazement across that field of waste. X.Y. began to smash one after another saying, "Perhaps I can find some of my tax money in one of these." Hundreds of luscious melons in that field and as many more in two other fields which we crossed nearby, dulled our enthusiasm for a time. We were sickened at the sight of such waste, especially since the price had prohibited our enjoyment of the fruits all through the summer.

Even though swollen with rage, we squeezed through a wire fence and again entered the woods. We had been skirting around a dried-up swamp. X.Y. was ahead of me going down to lower areas of a circular depression, when I heard the familiar war whoop which I knew to mean that a new plant had been sighted. I ran down the slope as fast as the Blackberry thicket would allow to witness the cause of the outcry.

"Do you see what I see?" he asked.

"I think I do. You mean that thicket of what looks like Spice Bushes that aren't quite Spice Bushes?"

"They are Spice Bushes of a different species," he beamed. "This is really rare."

Large, scarlet fruits in abundance covered low-growing shrubs of the compact colony carpeting most of the swamp. The larger fruits, thinner drooping foliage and lower stature showed at a glance that it was not the common Spice Bush (*Lindera Benzoin*) which grew on higher ground nearby.

"This is a rare Spice Bush (*Lindera melissaefolium*), of that I am certain. So far as I know it's one of the rarest shrubs in the United States. It was described in 1788, I think, by Walter in 'Flora Caroliniana.' He found it in swamps near his home in South Carolina. Other observations have been mentioned from Georgia, Florida, Alabama, and Louisiana." X.Y. had been sitting on the ground as if exhausted from shock.

Turning a branch over and over he continued, "What a skip from the Coastal Plain and Piedmont regions to here—some 400 miles from the nearest known locality, Wilcox, Alabama. Only two other collections have been made in the last hundred years." He poured out its history as if to fix this region in its proper space before it might vanish.

"How do you account for this berry bouncing 400 miles? It wasn't dropped here from a plane." From long association I knew beforehand what his next announcement would be.

"This means a trip back here in spring to get flowering specimens," X.Y. announced, as I expected.

"In hip boots then," I reminded him.

We took Kodachromes of the plant and left the area. Fortunately we emerged from the woods at the tidy, modern home of Mr. H., a farmer, who told us that this particular Spice Bush had been common since he was a boy. They called them "Pond Berries," and used them for shooting pop guns which they made from hollow Elderberry stems.

"That answers my question," I said. "Some lad in Alabama shot a berry into the air, and it landed in Missouri and no one knew where until today."

Nothing lifts X.Y.'s spirits so much as a new find—a strong stimulus to put wings to his heels. We went on down the hot sandy lane forgetting heat and sweat until thirst revived the memory of those fields of luscious melons by now collapsed as their frozen juices had melted under the sun.

While in the Spice Bush area, X.Y. had found a green tree frog (*Hyla cinerea*) which now rode folded in cool green leaves in his shirt pocket. This is the largest of the tree frogs though more slender than other of the North American tree frogs. This pea green frog is adorned with a band of gold extending from the front of the jaw back along the sides to the thigh. Gold spots on his back and golden eyes finish his rich attire. From the leaves of plants near lakes and swamps he can leap a distance of from eight to ten feet for an insect he may desire. He is sometimes called the "Bell Frog," because the notes resemble the tone of a cowbell.

Less than a mile from the swamp we came to a farmhouse to see if we could buy a melon that perhaps they had stored before the frost. They sold us three from a truckload they were taking to the hogs. When asked why the melons were allowed to freeze, the answer was, "No market."

"No market," X.Y. gasped, "I've been in the market for one every few days all summer, but couldn't reach the price."

There was still enough sunlight when we reached the car to photograph the green frog. X.Y. placed him on a sawed-off Cypress stump in sunlight not to his liking. Three times he hopped off before I could get the right pose. I finished the roll with several views of Cypress trees in fall shades of sepia—some upper limbs bending with fruits.

From here we drove northeast past Wapapello Dam and Lake on St. Francis River, over isolated roads up and downhill looking for a place to camp for the night. It looked as if we were to have another dry camp when at the foot of a long, steep hill an old well box of weathered wood appeared by the roadside.

"Dry," I said.

We stopped under large, virgin White Oak trees. "Dry or not we camp here," I said. The sun had already sunk behind the hills until not even the tips of treetops on the opposite hill showed a tint of sun.

From the limb of a small tree X.Y. picked a long, slender pole with a nail driven near one end as a hook. He hooked the rusty, half-gallon bucket found by the well box on the nail and lowered it into the rock-lined well. "Water," he said, "I've struck water."

"Its looks clean," I observed. "After boiling, it will do for dishwashing and bathing. There's no lack of hardwood here. After all our misgivings about a place to camp we've found about all in one spot—water, wood, a beautiful place completely isolated at a dead-end road, balmy and calm."

I sat on an exposed root of the old Oak which flung its branches to the north over the road, to the south over the well, and to the west over the car.

There I prepared the ingredients for the Spanish rice dinner. While the rice cooked, I carried the bits of refuse from the preparation to a distance as was our custom to prevent animals, especially skunks, from disturbing our sleep. We always tried to leave our campsite as clean or cleaner than when we found it, often burning or burying old trash left by former picnickers.

After we had finished eating, X.Y. tried to draw water again and lost the bucket in the well. I came to his aid with a flashlight. Fortunately the well was shallow enough to reach the bottom. Before we left the next morning he had developed real skill in sinking a bucket into the water while keeping the pail on the nail.

After breakfast we drove to a house on top of a hill near Hall's Bluff. There, a big man with a lame back directed us over to an old deserted house near the bluff. X.Y. was looking for a species of Alum Root (*Heuchera missouriensis*), a rare one which grew on the face of the bluff. It is known from only two stations in southeastern Missouri. This station was soon to be lost. The bluff, once shaded by a dense growth of trees and shrubs, was now exposed to full sun. A few bleached skeletons of tree trunks still stood. European weeds such as Ragweed, Plantains, Fleabane, and Mullein covered the talus slope below the bluffs where wildflowers and ferns were once abundant. Fluctuating water levels had laid waste to this section. A few of the rare *Heuchera missouriensis* still clung high where it was shaded by parts of the bluff. These would be drowned out in the future by high water levels. What a sorry contrast this area now represented with its former undisturbed state.

When it began to sprinkle I gave the return signal to X.Y. who gave me a bundle of plants to carry back to the car before they would get soaked with the coming rain. I started back over the hilltops. The "High Road" is usually much easier to travel; less underbrush, vines, and loose rocks. By following the ridge as closely as possible, and by skirting around the heads of ravines, for the most part, there is less climbing up and sliding down. I had learned that by such lines of least resistance I could return in a much shorter time, and seldom got lost. Here a sense of direction is important in order to return to the car or somewhere up or down the road not far from it. To go through a trackless Ozark forest and come out at the desired point one needs more than the built in sense of smell for ants on the trail. I once removed a leaf six inches long which formed a part of their scout-prepared trail to their source of food, consisting of oranges and bananas on the bird feeder. At that spot the ants both coming and going were completely bewildered, wandering in every direction for more than one-half hour before one stumbled on the far end of the trail broken by removal of the leaf. The scent laid on that leaf by the scout did not penetrate through the leaf to the ground, and their sense for picking up the trail did not span the distance of six inches. Smart ants? No. Efficient robots? Yes. At any rate, this time I was smarter than the ants and beat the rain. I sat in the car dry, and read until X.Y. came in dripping from head-to-foot.

From there we drove to Wappapello Dam and beyond, where X.Y. looked over the lowlands for Rose Turtlehead and found none. Later at Stanley Creek we obtained permission to camp on a spring branch by some long picnic tables. Darkness came before the car was unloaded. Table de luxe—after standing on my head day after day preparing food and washing dishes, this old table of rotted planks was indeed a luxury so long as I did not set something where a plank was missing. I could stand upright and prepare the hunter's stew. The table was great, but unfortunately the "kitchen" had no roof. Rain, though not hard, fell enough to threaten the fire which I had to keep coaxing. We agreed that the stew cooked over the smoky fire absorbed enough hickory smoke to flavor it quite properly. While eating in the drizzle our damp clothes took on some of the hickory fragrance as we stood around the smoky fire.

Even though the rain had ceased soon after we had eaten, I prepared my bed in the car in case it should come again. X.Y. disdained the car. He wanted fresh air. Sometime in the middle of the night I was awakened by his, "Woof, Woof!" I raised up to see what had brought about the outburst. In the bright moonlight I could see him sitting up throwing flint rocks at the farmer's hogs, which had him surrounded. For a full 15 minutes flint rocks hit flint rocks or live sides of bacon, knocking sparks of fire or squeals of pain, as the case might be. Finally, the hogs became discouraged and gave up their curiosity about the dark bundle lying among the white chert rocks.

At five-thirty I got up to make hotcakes, the breakfast we found to be most sustaining on long, strenuous hikes in cool weather. X.Y. left at seven to hike through wet bushes down Stanley Creek. After camp duties I rambled over hills of pink, white, and lavender chert, too barren for much plant life. I was searching the dry bed of a little gully for a hollowed flint rock to use for a pool in our terrarium at home, when I saw X.Y. coming followed by a brown-and-white spotted dog. At last X.Y. had found what he had been hoping for, a swampy meadow with broad-leafed Rose Turtlehead, and the narrow-leafed White Turtlehead, Closed Gentians, and a swamp species of Lousewort. The Turtleheads, especially the rose one, had been his special aim.

While on the hike, carrying a bunch of Turtleheads, X.Y. had stopped at an isolated farmhouse where a woman was doing the family washing. He introduced himself by saying that he was studying the flowers of that region. She answered that she did not want to buy any flowers. It was always difficult to put across the idea that he was merely studying them. He showed her the purple ones to see if she might remember if both white and purple ones grew in the same swamp, but she knew nothing of the swamp, having recently moved there from an area already flooded by the dam. She assured him that he was welcome to anything he could find in the swamp, and the dog, which did not belong to anyone she knew.

At eleven-thirty we filled our containers with the unusual "soft" spring water and left with the dog running up the hill ahead of the car at 15 miles an hour. For two more miles he followed trying hard to keep up on level road. With our increased speed he lagged farther and farther behind until we lost sight of him.

Near Dexter, fall colors furnished subjects for Kodachromes. Here the green of the Huckleberry was touched with red-purple; Willow Oaks were mixed with yellow-and-green; leaves of Basket Oak were margined with rose-and-maroon; most Cypress had turned to rusty orange. I finished a roll with snaps of cotton pickers in the field and was sorry I had no way to accompany the scenes with the songs of mockingbirds that were so much a part of the setting.

Near the southern part of Pemiscot County we stopped at a filling station to inquire about a road that led to a spot X.Y. wanted to explore. The man insisted we could not get there because of too much water. Also he warned us that there were bears in the woods along the bayous of the Mississippi River.

We drove southward along the bayou jungle, looking for a place to camp. On each side of the narrow one-way road the unbroken, dark and impenetrable swamp jungle pushed up to the very edge of the road. Signs along the edge of the road warned not to pull off the road because of soft shoulders. Dusk had turned slaty and the road which led only to swamp dwellers, gave no promise of a change for the better. There was no choice but to back up the entire way to an open place I had noticed by a large tree. There we found space with enough solid ground to pull off the road. I scouted for a few bits of wood to cook our supper of soup, steak, and potatoes. With my flashlight, I tramped around in the small open area trying to find a spot free of Poison Ivy where I could set up my cot. By tucking the covers firmly in at the foot to keep them from brushing the ivy, I could have a poison-free spot near my head.

We sat on the sandy soil in a scoop left long ago by the flooded river, and ate our supper by the light of the camp fire. The small spot which only a short time ago had appeared so dark, foreboding, and perilous as we passed, now, with familiar articles of housekeeping spread about, in the light of the camp fire, seemed bright, cozy, and safe. We sat, each with a half of watermelon on our laps, enjoying its chilly sweetness when a noise crashed through the bushes behind us.

"Bears," I said, "The bears the man at the filling station told us about."

We were not anxious to lose our food or to argue with a bear about who would get it. X.Y. put more wood on the fire, and we continued to eat listening to the crashing sounds now nearer, then farther.

"It could be a person curious about our fire," X.Y. suggested.

We felt slightly uneasy at this unfamiliar sound. No sooner would we think the source of the sound had gone than it would occur closer or from the other side of the road. Were they closing in on us, we wondered? The "sound" or the "thing" had aroused our curiosity to say the least. There seemed excuse enough, after finishing the melon, to take the rinds out along a narrow strip of sand to dispose of them and incidentally, with the aid of the flashlight, try to determine the source of the sound. I sat by the fire, having had experience with bears in national parks, while X.Y. disappeared into the darkness. The sound had been coming mostly from our side of the road. There was a brief sense of relief when I heard the noise at a distance on my left, then within seconds a crash came from the very edge of our camp. The whole bear family, I thought, relieved to see the circle of flashlight bobbing along toward camp.

X.Y. came in smiling. "I have the bear behind me," he announced, then tossed a Hedge Apple—Osage Orange, as large as a grapefruit at my feet. "Here's the bear," he said. Frost of the preceding night had loosened the apples which were crashing down through dead branches and brush under the trees.

X.Y. placed more Osage Orange wood on the fire, sat down, and related its merits. Osage Orange trees, native to the land of the Osage Indians in Missouri, furnished the Indians and early white settlers with hunting bows. For this reason early French settlers gave the name, "Bois d'arc" or bow wood. In early 1800, a good Osage Orange bow could be traded for a horse and blanket. Some archers still prefer bows made from this wood. The tree is passing from its one-time popularity as windbreak, fence, boundary, and fence posts which lasted up to 40 years. Incidentally it served well for a wildlife sanctuary. Then came fence wire, and a desire for more space weed clean right up to the fence; and tractors with tires subject to punctures by the strong spines of the Osage Orange limbs. Whole rows of trees were pulled out and left lying with their orange-yellow roots, once used for a yellow dye now exposed to weathering. Game diminished, birds had to move to more wooded quarters, and insects increased to plague the farmers. At present the fashion is back to live boundary, the Multiflora Rose fence. You can see Osage Orange is good camp fire wood, starts easily, is smokeless, and makes good coals," X.Y. added.

"What are the fruits good for, besides to scare people?" I asked.

"That remains for some chemist with enough curiosity and perseverance to carry it through. They have extracted an antioxidative which may prove to be useful in preventing oily substances from becoming rancid," he answered.

We went to bed as a bright moon came up and lit our area all night. The familiar sound of the screech owl, shrieks of large water birds, and two flocks of Canada geese disturbed us little. Even now the crash of the Osage Orange had become a familiar sound.

The quaver of the screech owl did not send us into panic as it is known to do with two other races of people. The Germans in one known settlement in Illinois kill the screech owl that comes near the house because of an old belief that if one makes its weird call some member of the family will die. How did that belief span continents, thousands of miles apart from Germany to South America, where the Indians of Colombia believe the same?

By daylight the next morning we gathered more wood to cook our breakfast of ham, eggs, and coffee after an orange appetizer. As we ate breakfast a huge flock of bronzed grackle, two lone geese possibly lost from the flock that passed in the night, and another noisy flock of geese flew southward reminding us that we, too, must start homeward. X.Y. gave up trying to reach the flooded section. We left camp at seven-thirty returning northward through Hayti and Cape Girardeau. The early start would put us home in daylight, hence the question, "What will we look at first: the wildflower woods, the fall flowers of the prairie, or the cultivated section?" This thought supplied the transition from the primitive life we had just finished to our more everyday, refined existence.

CHAPTER XVII

OCTOBER

Camelhump Exhibit of Fall Colors

There had been frosts and freezing nights at our home in Barrington before we set out for the last attempt to augment the "Flora." It was nearing the last of October, and we had reluctantly released into the cold our crippled hermit thrush which I had housed for over a year. Now instead of sitting on the back of a chair in the kitchen singing with the steaming teakettle after helping himself to a breakfast of specially prepared foods for thrushes, it was heartbreaking to see him sitting on a frosty high cranberry bush eating the frozen fruits which were as hard as rocks.

Since giving him his liberty, he still chirped at me from shrubs near the edge of the woods; and he still came near the house to bathe in his favorite black bakelite plant saucer, his own choice. Now half domestic and half wild, he was on his own in the wild.

We kept wondering how he would make it southward with a crippled wing. The time needed to traverse the long highway from Barrington to St. Louis, usually considered monotonous, went more swiftly as we planned the hermit's route. All along Highway 66 we noted woodlands, patches of shrubs and places where a thrush could spend the night. It seemed possible that he could fly from one wooded area to another in one day even with a crippled wing. We took some comfort in the possibility that he could possibly reach the Mississippi flyway southward before winter became too severe.

This had been a very dry summer and in addition to looking for plants for the "Flora" we intended to take note of the effect of the drought on trees throughout our journeys in Missouri. Also it could be a very good year for rich fall colors. Three factors contribute to good fall colors: a fairly dry fall; warm bright fall days when leaves make much sugar; followed by cool nights between 35 and 45 degrees. Under these conditions sugar is trapped in the leaf, especially in the part of the tree most exposed to light. The reds caused by an anthocyanin pigment produced by an accumulation of sugar in the leaves of Maples, and by tannins in the leaves of Oaks, appear when the chlorophyll breaks down exposing the yellows and reds present in the leaves.

Therefore, when the temperature falls below 45 degrees, it is conducive to the orderly closing down of tree factories for the winter, and we have in the leaf, the yellows and reds left for further chemical changes.

With Camelhump in mind as the best lookout to view fall colors, we stopped for the night on Crooked Creek near Sligo. It was a quiet night but an exciting breakfast. While we were eating on the very edge of the clear creek, a deer ran full speed out of the bushes across the stream close enough to splash us with water as he dashed through. It would be difficult to say if the deer, or if we were more surprised. Probably we were trespassing on his usual run since we were some distance from the road.

The visit by the deer, even though only a quick splash from slender hoofs and an instant flash of a white tail, put us in a pleasant mood for the drive along the crystal clear Huzzah River where uncultivated bottomland fields flaunted banks of smooth Goldenrod (*Solidago altissima*), the most beautiful of all Goldenrods. Standing three feet tall, they caught the morning sun effusing a golden glow broken by amethyst spots of starry Asters, loose sprays of big blue Asters, and dense racemes of small pale violet or white Asters. Untrimmed roadsides now exchanged their former weedy border for ribbons of yellow, blue, rose, and white as we drove swiftly by. Old rocky roadsides that seemed too poor to produce anything but weeds, were now enriched with unsuspected colors. We were happy that the weeds had not been cut.

In late afternoon we reached the crest of Camelhump, an open knob scorched by summer heat, and swept by winter blasts. Clones of Persimmons filling spots here and there on the hump exhibited fruits already changing from orange to powdered blue, though most of them were solid and quite capable of caustic effects.

From the hump higher than the surrounding country, one can look across miles of valleys and varying heights of hills fading in the distance. To the north and east as the sun broke through, the late afternoon clouds turned a giant spotlight on the opposite valleys and hills, causing the colors to burst into brightness. Along the creeks, soft yellows marked the border of Willows. Above them the Silver Maple's yellow, and Red Maple's scarlet stood apart, clear and distinct from the more numerous Sugar Maples which dared to run, in a single leaf, the whole gamut of rainbow colors minus blue.

Dotted over hills and slopes the clear yellows of Hickory and Locust, and the rich blue-green of Laurel Oak and Black Jack Oak blotched the dominant color red. Nature had strung her whole series of red, with every overtone of color, as far as the eye could see, from the firey red of the Scarlet Oak through the deep reds of Shumard's Oak and purple-red of Sour Gum to the soft rose-red of flowering Dogwood.

As the clouds shifted past the sun, one hilltop after another was shot with light, intensifying the colors in contrast with the hollows already beginning to take on the purple tints of dusk.

We sat spellbound gazing out over colored hills repeating themselves like Ravel's "Bolero" on reverse, drumming on endlessly with slight variations, swelling in color as they neared a crest and fading away with distance, and diminishing daylight as they descended into the lowlands, but never at any time monotonous. We were tempted to sit until the highest crest turned purple, but we had to hurry on to find a place to camp.

Devil's Run Creek, a very rocky, wild, clear creek was a nice place to camp by an old swinging bridge gone to ruin. There we slept under golden colored Maples, with hot flint rocks for warmth as it froze that night. Freezing temperatures had silenced all of our usual serenaders, even the big hoot owls. X.Y. barely finished saying, "What has happened to the owls?" when a screech owl piped up as if he had heard the remark. I was surprised to see glow worms while we ate supper.

At dusk one kingfisher squawked up the creek almost brushing the water under the old swinging bridge. At daylight he squawked back down the creek again under the bridge. Considering the speed at which he passed through the narrow space between the sagging bridge and the water, I expected him to crash, though his precise aim was no surprise since his eyesight is supposed to equal that of the eagle.

Hotcakes for breakfast by a camp fire were enjoyed more, especially since the collecting year was drawing to a close. The clear still pool backed by a moss- and fern-covered bluff caught and mirrored the golden Maple leaves slowly dropping from limbs overhead. No more beautiful site could be found for a hotcake breakfast. Royal Ferns here also added a bit more charm to the place.

We had entered the state earlier on Highway 66 southwest of St. Louis. We had seen that Elms, Oaks, evergreens, and many other trees were dying where the woods were overgrazed, whereas trees on other equally rocky slopes nearby covered by underbrush and heavy leaf mulch, were in good shape. The scene was repeated beyond Mountain Grove on Highway 60 where grazing had skinned the hills to bare rocks. It looked as if the animals had eaten the soil from the slopes. Herds of goats showed up every few miles making a desert of acres and acres. Where overgrazing had cleaned off every bit of vegetation, stock still tried to find a green blade. Trees at such places were covered with dead and dried leaves or had lost their leaves prematurely, while just across the road where no grazing had been done for years and leaf mulch and undercover remained, the trees seemed to take the drought fairly well.

Beyond Willow Spring we saw a sign, "Goats for sale. They do a clean up job."

We had no doubt about that. They left only the rocks in their wake. It would be difficult to imagine any cover over those rocks in the near future.

We continued on Highway 60 to Mountain Grove where we visited sinkholes and natural upland ponds. One sink was about one-fourth of a mile wide at the top with bluffs on the sides that dropped abruptly down about 100 feet. Very likely the roof of a cave had given way many years before. We did not go down into that one.

Near Beaver Creek I tried to take a picture of another swinging bridge, but could not get the right angle because of dense masses of Poison Ivy guards at each end. It was probably just as well since the fashion of swinging bridges had long since passed and the bridges were no longer safe for venturesome feet. In the evening while we were there a farm boy came by leading his hound dog, ''Pearl,'' enroute to a neighbors to join a hunting party for the night.

According to the names of creeks and towns, this country must have at one time had larger game than the present possums and coons which furnished the local people with nighttime sports. At Elk Creek, a clear spring had abundant Water Cress, but at Jerktail we found the country dry and overgrazed.

We left this dry region and drove to Bennett Spring State Park. It rained all night while we were safe under shelter. In the morning where we stopped to take pictures of tree-covered bluffs in their fall colors along the Niangua River, X.Y. saw a black coach whip snake about four feet long at the base of a rocky bluff. Two-thirds of the way up on the southwest-facing side of the bluff he saw a black widow spider on a web in full sun. This was less exciting than the one he once found in his sleeping bag. What were snakes and black widows doing out so late in the fall?

We returned to Bennett Spring to camp again that night. There was a cold drizzle. Fog rose like smoke from the warmer water of the spring branch. At daylight we heard a siren. Cars and people began to run. One man in hip boots appeared, then others rushed in. A trout rodeo was soon evident. The siren was the signal to start fishing for the largest trout. Each numbered man waded into the water of the spring, which was really warmer than the air.

We did not stay to see who caught the prize trout. After loading our car we left, going eastward through the bayou country down to Hayti and Caruthersville. At Bell's Point the scenery changes abruptly from the Ozark regions. We were now in the land of cotton fields, cotton gins, soybeans, and levees. I watched X.Y. walk out across a flat soybean field followed by four mules.

Later when we stopped on a road to look at some woods, everyone came out from his little home to ask what we were selling. When we stopped near Naylor we were asked to buy. Molasses making was in progress. We passed one mill where the juice was being crushed from the cane and boiled in long open pans. One man using his best salesmanship came to our parked car to offer us a sample spoonful of green molasses barely cooled from the pan—green and thick, the very best.

On the way home toward St. Louis, X.Y. stopped to look at a swamp in Reynolds County, the only county in the state where the Snake-mouth orchid (*Pogonia ophioglossoides*) had been found. The swamp had been rooted up by hogs. No sign of leaves or stalk could be found. One eleven-year-old girl was much concerned about the orchids having been destroyed by hogs and cows. She showed us the fall Eastern Witch Hazel in bloom along the spring branch above the house, but that was weak balm for the gash left by the hogs in the only known *Pogonia* colony in the state. X.Y. left with a tortured vision of that rooted-up *Pogonia* swamp.

From appearances, the area had once been a swampy valley covered with the usual shrubs and trees for that region, sheltering the unusual Pogonia. It had been cleared for fields with a hog lot in the swampy area below the spring. The hogs had spared nothing as they transformed the rare site into a muddy wallow that had known no green save for algal slime for many years.

This October we had seen fall colors in mass and many striking individuals.

Those predominantly yellow were:

Ailanthus–pale yellow with green,
Vitis–pale yellow with green,
Osage Orange–pale yellow with green,
Catalpa–pale yellow,
Honey Locust–rich yellow,
Silver Maple–pale yellow with green,
Black Willow–pale yellow with green,
Tamarisk–rich yellow with green,
Pawpaw–yellow
Willow Oak–yellow with green,
Winged Elm–yellow to orange with green,
Black Oak–burnt yellow, tan, and green,
Pignut Hickory–clear yellow,
Ward's Willow–lemon yellow with green,
Black Walnut–bright yellow,
Butternut–bright yellow,
Chestnut Oak–yellow with green,
Slippery Elm–pale yellow,
Beech–yellow to orange,

River Birch–pale yellow to russet brown with green,
Black Jack Oak–variable, old gold with dull green, dull yellow tinged with orange, dull orange with reddish salmon, and burnt orange,

Those predominantly orange and red were:

Sassafras–orange to red,
Dogwood–rose to red,
Viburnum rufidulum–deep rose-red,
Sugar Maple–all shades of yellow, orange, red, and green,
Sour Gum–red to purple-red,
Red Maple–scarlet red,
Red Oak–red,
Basket Oak–dull green with margins of rose or red-maroon,
Pin Oak–grass green with dull reddish yellow,
Scarlet Oak–scarlet,
White Oak–rose, red to purplish-red,
Post Oak–variable, yellow, dull orange, rose, green, and russet brown,

Those predominantly rusty to brown:

Cypress–burnt orange to brown,
Sweet Gum–all colors sometimes on a single tree even to purple at different heights on the tree,
Shingle Oaks–remained green to this time.

In general the trees, shrubs, and herbs of the Ozarks had not suffered too much from the drought except where overgrazed or burned. There, the trees either had dead leaves, or the trees had died entirely.

The month of October had passed, yet a few more places remained to be seen. Near Davisville we spent a few hours at the stone house where we took pictures and picked up blossom rocks for our grounds at home. Then we drove to Round Spring. Late in the evening we visited Round Spring Cavern. Near the entrance, the cavern branches going in different directions. Following the left-hand trail along a stream one sees: the exit of a stream apparently through solid rock walls; the usual bat "hangout"; the "tobacco barn"; and a 14 foot waterfall which blocks any further passage in that direction. On the right-hand trail we came to the grinning "Dragon's Mouth" blocking that exit. Huge open jaws each containing hundreds of sharp teeth warn all that no pass is permissable. Another very interesting feature on this trail is the "Drape," a curtain type of stalactite resembling folds of hanging cloth. Illuminated from the opposite side, the translucent drape shows stripes and designs due to the different colors and thicknesses of the calcite drape. Since there was no place at Round Spring Cavern open to accomodate us that night, we drove on to Alley Spring, arriving in the rain.

At the shelter house a teacher was giving lessons in square dancing to a mixed group of all ages. The guitar and calling continued into the cold night. While the youngsters puffed from vigorous exercise, we, on the sidelines, shivered in our heavy jackets. At midnight we left them still going around and around in the big shelter room.

The next morning when I took my first look at Alley Spring, it was going around in circles. A migraine headache had taken up where the square dance left off.

Along Highways 106 and 49 we spent some time taking photographs of colored trees. At Oates Conservation Camp in Clark National Forest we climbed up the lookout tower where the views of the colored hills were masked by mist and the valley marked by fog rising into the cold November air from rivers and creeks. Where was Indian Summer?

The night was spent in an isolated rustic shelter. For privacy, convenience, and comfort, this is a fitting retreat for naturalists. A road shut in at the foot of a long steep hill presaged that the only people who would come, would be those who were intending to camp, and November is no idle month to camp. Plenty of dry, cured hardwood already cut, was handy at the huge stone fireplace. Just outside the door within two steps was clean uncontaminated running water. Uncontaminated air also put zest into life on this late fall visit. On the creek banks, the fall-flowering or Eastern Witch Hazel (*Hamamelis virginiana*) was in bloom. This is about the last plant to bloom in the fall, while the Ozark Witch Hazel is the first to bloom in the spring. Another difference in the two species is the variance in the length and color of petals. In the fall-flowering one, the petals are longer and a clear yellow, while the shorter petals of the Ozark one are yellow-orange with a touch of rusty-red. Also the fall-flowering one is found in a cluster in only five counties in the Iron County region, while the Ozark spring-flowering one is spread throughout the Ozarks.

We could not leave the Ozarks without a fall visit to the "Johnson Shut-ins" where we hoped to get some good Kodachrome shots. The sun failed to come out, however the mist was just right to put a soft touch on the color views of the Red Maples and evergreens, and the rose-to-purple granite and gray rocks forming the background of the clear, blue water. The view exceeded in charm those that might have been taken in strong sunlight—a fitting climax to the close of an exploring season.

CHAPTER XVIII

ADVENTURES

The first 17 chapters represent only a few experiences, a mere peep behind the scenes, of our journeys together during the 30 years of research on the "Flora." Many times X.Y. went alone. At times he was accompanied by other botanists or friends. The following adventures were experienced by X.Y. when I was not along.

On one September weekend trip alone he went to locate, by means of an air photograph, a salt lake hidden at the base of a hill along Elk Lick near Paris in Ralls County. No one was at home on the farm to give permission or assistance for finding the place. After some time he found the lake, but the only saline plant there was a small Spike Rush (*Eleocharis parvula*). This trip was not in vain, since this established the third location known for that plant in Missouri.

From there he went to Mark Twain State Park to botanize the bluffs, and found that the park had been kept natural and wild. In the park he picked up a companion for the rest of day, which he thought was a blood hound. A friendly dog could very well be a good companion when rambling alone in the wilds where no one would be missed or found for weeks.

He also profited by being alone in the shelter house at supper. A party of picnicking teachers noticed that X.Y.'s supper consisted of a can of chicken noodle soup and one of creamed tomato soup. He must have hidden his cakes or ever-present cookies when he saw their chocolate cake, about the most irresistible dessert, short of a chocolate malt, on his list of choices. They donated three slices of the cake and two of watermelon.

The party left at nine o'clock. The dog had gone home at dark and X.Y. was left alone at the shelter house with the katydids. Even the familiar barred owl and whip-poor-will failed to question or command.

The next morning he again profited by the picnickers' coals from the fire of the night before. These were handy for cooking an egg omelet and cocoa in order to get an early start revisiting a low wooded area along Salt River on Highway 15 north of Shelbina. At that place he and Palmer had found the only known locality in Missouri for *Viburnum pubescens* var. *Deamii*.

On the unbeaten trails of rural Missouri, the study of flowers was the least likely occupation or pastime a man could be expected to follow, therefore it was not surprising that X.Y. was mistaken for an herb doctor, an artist, a plumber, or an Indian at various places in the state.

When stopping at farmhouses for permission to botanize over their area, X.Y. usually began with, "I am making a study of the plants in this area. . . ."

Immediately they identified him as an herb specialist or a doctor and requested help for some ailment. It always ended with an exchange of herblore.

At other times he was mistaken for an artist as he entered with his plant press slung over the shoulder. Once he was thought to be the paperboy.

The worst mistake was made when a farm woman seeing him walking down the railroad track called out, "The house is up here."

He called back, "I am not looking for a house."

Then she asked, "Aren't you the plumber I called for?"

X.Y. and one assistant almost beguiled themselves into believing they were rainmakers. That was a record year for drought all over the Ozarks. There had been no rain for months until X.Y. and his companion entered a county. On that day it would rain. As they moved from county to county rain followed so regularly that they began to believe they held the charm, and when a farmer would complain that no rain had fallen for weeks, X.Y. would promise, "Don't worry. You will get rain today, because we have entered the county." Rain followed them all across the Ozarks, coming as they entered each new county as if the rain was timed with their entry.

Early in his collecting experience one July, X.Y. wanted to explore the bluffs along Osage River in Miller County. He went to get the owner's permission. The man agreed in exchange for a favor. Obviously, the man with a certain sense of humor, quickly seized upon a scheme to put it to use. X.Y. being quite dark skinned by nature and still darker by a summer of exposure to the sun, and further, somewhat in need of a haircut after an extended time in the wilds, suggested "Indian."

The owner had in his employ a "George" who was eccentric on the subject of Indian legends in this part of the country. He also believed there was gold hidden in the hills.

On the side, the owner said to X.Y., "Your skin is quite coppery looking, maybe you can play the part of an Indian."

When X.Y. consented to play the part, the owner called George over from the work he was doing and said, "I want you to meet an Indian who has just come up from Oklahoma."

George, who was about 30 years old, fell for the joke. His eyes popped, "What part are you from?" he asked.

X.Y. thought fast and answered, "Muskogee."

George beamed, "Yes sir, that's the right place." Then added, "What's your name?"

X.Y. had not thought of a name, "Star Mark," sounded like a good Indian name, and the first he could think of.

"A good name," said George, then asked, "What are you here for?"

Looking far away toward the bluffs in the distance, Indian fashion, X.Y. whispered mysteriously, "To explore those bluffs over there."

Then George answered his own question, "I think I know what you're here for."

"You do?" X.Y. said eagerly.

"Yes, I know of hidden treasures around here. I can take you to them. I know all the best places for looking."

As X.Y. did not wish to be delayed or to reveal his real purpose to George, he was faced with inventing another story. At that time he was using the huge green vasculum especially made for him. He told George he would make a quick survey and return before he could finish the job he was working on at the moment.

When George saw X.Y. returning after the plant survey he hurried forward with excitement, "Did you find it?"

"Oh, I just surveyed the country, and I'll come back later," X.Y. answered casually as if hunting for treasures was his daily business.

"What's in your can?" George fingered the can suspiciously.

"Only some plants I picked up on the road," an honest answer.

"I'm a good guide. I know all the trails and landmarks, and when you come back we'll go together looking for the gold left by the Osage Indians during the Spanish invasion."

The gold of the Osage Indians is still in Osage County unless George found it.

While others made mistakes in the identity of the plant collector, he himself sometimes failed to recognize the images that lurked around the corner of his daily proceedings. Adventure usually admits that the imagination is not capable of sensing that which awaits the one who dares to venture carelessly beyond his immediate vision. The experimenter is an adventurer always aware of the unexpected just out of sight. If he is wise, he keeps one eye on the experiment and the other eye around the corner.

X.Y. wanted to try an experiment. Drying plants with blotters which had to be spread in the sun was not always the most reliable procedure. There were times when rain, several days in succession prevented drying the blotters. Therefore, for a long time he had been figuring out a device for drying plants on rainy days. He had a wooden box made sufficiently large to hold plant presses. These presses could be supported over small lamps under the box. Corrugated pasteboard sheets cut the size of the blotters and placed between specimens permitted warm air to circulate between the specimens. The idea was good and worked well for awhile.

At that time he was using a panel pickup truck for exploring and the plant drier fitted easily into the truck. One time, when two assistants were helping him, someone turned the lamp on too high. X.Y. saw smoke coming out of the car. All three jumped into the car and extinguished the flames, drenching the presses with water but saving the car. Some specimens were lost and others scorched, though the burns X.Y. received on his hands and arms prevented his recording, which had to be carried on for several days by his assistants. That ended forever his plant drying in the car.

At another time when wearing high rubber boots he posed by a camp fire for a three minute time exposure. The camp fire was blazing high and hot for the picture, and X.Y. was too close for comfort. After about two minutes he told the photographer he was getting too warm. "Only one more minute," the operator said.

That was one minute too long for rubber against the leg.

He took a swim immediately afterwards, followed the next day by a long hike to Tom Sauk Mountain. Pains were shooting up the burned leg before he reached his destination. There still remained the two miles back to the car and the agonizing drive to St. Louis where the doctor pronounced it a first degree burn, drained the infected leg, and put him to bed. A permanent scar remained to remind him of the price he paid for a time exposure.

On a trip alone in north Missouri X.Y. was driving downhill on a country road and shut his eyes to that image around the corner, or more properly did not take time to admit consciously or otherwise that anything could happen other than what was immediately on his mind.

He spotted a Blazing Star, stopped the car, and without putting on a brake, or putting the car in gear, he jumped out. While getting the flower he saw the car starting to roll toward a steep gully on the right. He jumped into the car and pulled on the emergency brake just as the right wheel was going over the edge of the road. The rear right wheel still hung on the dirt, while the front right was in the air.

With a spade X.Y. shoveled a mountain of dirt under the two wheels until he thought the car would stand there while he went for help. No one was at home at the first two places he tried for help. Finally after much more walking he found a man who would come with chains to pull him out.

Probably the most exciting adventure in all his 30 or more years of exploring and living outdoors in Missouri, was not one due to any lack of foresight on his part. Wild animals usually live peacefully with naturalists who give the animals their proper degree of respect. However, in this case it was otherwise.

In the fall while on a collecting trip along the Meramec River with an assistant, X.Y. put up a tent as the night was cold and rainy. They also stacked firewood under the car to keep it dry for the morning.

In the middle of the night the assistant called out from his sleeping bag on the ground, "Ouch, my finger. It's got my finger."

Startled out of sleep, X.Y. at first thought a snake or turtle had struck his companion. Almost before wide awake he reached over in that direction and the same thing grabbed and bit his finger. When he turned on the flashlight there stood a skunk at bay. The men both shot out of their sleeping bags and out of the tent followed by the skunk. When they started to go into the car, the skunk began to walk away, looking back as he went. They then grabbed some large sticks of wood from under the car and hurled them at the skunk, finally killing him. They went to the river, washed their bitten fingers, and with no further treatment went back to their sleeping bags. The next morning they drove away leaving the dead skunk where they had killed it, and for two days more continued on their journey.

When at last they returned to St. Louis, they both took the Pasteur treatment of ten shots for rabies, though it was not possible to examine the skull of the skunk, which had been left in the woods.

They never knew if the skunk was mad, or if it merely wandered into the tent and naturally grabbed at moving hands in the dark, but unnaturally never sprayed.

That was X.Y.'s bad month. He had barely finished the shots for rabies when he had to begin tetanus shots after running a rusty nail through his tennis shoe while climbing over a fence near a barnyard.

It was a long exploration of more than 30 years from which we have related merely a few experiences representing each month of the collecting season. So we see—a book is not only written and rewritten, it is lived through a life of sweat, privations, and joys behind the scenes, before the first word is ever put on paper.

We, X.Y. and I, hereby will the state to the next botanist.

INDEX